Lathe Operation and Maintenance

John G. Edwards

Hanser Gardner Publications
Cincinnati, OH

Library of Congress Cataloging-in-Publication Data

Edwards, John, 1948-
   Lathe operation and maintenance / John G. Edwards.
      p. cm.
   ISBN 1-56990-340-9
 1. Lathes. I. Title.
   TJ1218 .E39 2002
   621.9' 42--dc21

                                                                                                                                             2002009338

While the advice and information in *Lathe Operation and Maintenance* are believed to be true, accurate, and reliable, neither the author nor the publisher can accept any legal responsibility for any errors, omissions, or damages that may arise out of the use of this advice and information. The author and publisher make no warranty of any kind, expressed or implied, with regard to the material contained in this work.

A Modern Machine Shop book published by
Gardner Publications, Metalworking's Premier Publisher
www.mmsonline.com

Hanser Gardner Publications
6915 Valley Avenue
Cincinnati, OH 45244-3029
www.hansergardner.com

Portions of this book originally appeared in *Modern Machine Shop's Handbook for the Metalworking Industries.* Copyright by Hanser Gardner Publications.

Copyright © 2003 by Hanser Gardner Publications. All rights reserved. No part of this book, or parts thereof, may be reproduced, stored in a retrieval system, or transmitted in any form or by any means without the express written consent of the publisher.

Table of Contents

Acknowledgements iv

Chapter 1 - Safety 1

Chapter 2 - Machine Maintenance 13

Chapter 3 - Engine Lathes 25

Chapter 4 - Toolholders 59

Chapter 5 - Cutting Tools 81

Chapter 6 - Workholding Devices 115

Chapter 7 – Part Alignment 159

Chapter 8 – Cutting Parameters and Tool Geometry 171

Chapter 9 – Cutting Fluids and Tool Cooling 197

Chapter 10 – Basic Cutting Procedures 211

Chapter 11 – Project Planning 245

Review Questions Answer Key 252

Index 257

Acknowledgements

The quality and completeness of any work is the sum total of those people, places, and entities who unselfishly give of their time and resources to make good things happen. This publication, once again, demonstrates that premise to the highest degree. A great deal of effort has been expended to bring the best of current thinking and industry-accepted practices from a community of friends, individuals, and machinists. It is a shared vision.

This book is written for the novice and experienced machinist-in-training. Everyone involved with this project has shared his or her years of experience to help shorten the learning curve. We don't want anyone coming away from reading the text to feel lost, so every effort has been made to clearly explain the principles of operating the lathe. Many examples, photographs, charts, and illustrations have been included to help visual learners.

This book would still be an idea and thoughtful musing without the help from a great many family members, friends, and companies. Their kind words and encouragement were often very welcome in the wee hours of the night. Without their generosity and wisdom, these pages would be empty indeed.

My sincere thanks and gratitude go to: Evaristo Aguilar "Jr." (my automotive protégé); Jennifer Blake (my sister) for providing additional encouragement along the way; Steve Hoppe of B&R Tool and Supply Company; Bridgeport Machine Company; Randy Buchannan of Buchannan Precision Machine; John Edwards (my dad); and Javier Garcia. And *Thank You* to Jeanie and Kenny Helderlein; all the staff at Helderlein Engineering Consultants for allowing me the invade their shop; Ralph LoStocco; Jack Luper at JEL-CO Precision Machining; Michael Lutfy (for having the courage and foresight to print my first book, the *Auto Dictionary*); Mark Magańa (my machinist protégé); Schneeberger Machine Tools; Sunnen Products Company; and John Van Zant (my best friend and mentor).

Safety

Safety in the shop is very important. You will not be able to perform the duties of a good machinist very well with only one eye, or missing fingers and limbs.

Always think about your clothing when working around machinery. Loose fitting clothes can get caught in machine parts, and lead to injury. Wear snug fitting pants and shirt, and steel-toed shoes or boots are recommended. Also, the less jewelry you wear, the better; bracelets, watches, chains, and even rings have all been known to cause injury.

Working around machinery requires a high awareness level of your surroundings and machine operational procedures. Using your senses to see, hear, and feel what you do may save untold pain and suffering. Keep your eyes on what you are doing. Listen for unusual sounds. Also, touch your lathe and learn to interpret its vibrations—you should be able to distinguish good vibrations from bad ones.

Look...

You are only issued one set of eyes when you are born. You must learn to take care of them. Wear eye protection whenever you work around tools and equipment. Your eyes are the most delicate part of your body. They are sensitive to flying objects like chips and broken tools.

Eye protection takes several forms: full-face shields, goggles, and glasses. Each has a place within the machine shop. Face shields utilize a large piece of thin plastic sheeting attached to an adjustable headband. When worn properly, they keep large pieces of debris from colliding with your face—a smart choice when turning a long workpiece.

Goggles are protective lenses that use either a rigid or flexible frame that fits snugly against your face. They are held in place with an elastic band. Goggles are especially recommended for use around grinders because they provide complete protection for your eyes from flying abrasive grit and metal shavings.

Glasses provide the most convenient form of eye protection and are, by far, the most comfortable. There are a variety of frame types, including all plastic, or metal frames with shatter-resistant plastic or glass lenses. The majority of safety glasses will incorporate some type of side protection.

Most machine shops and schools will have a minimum eye protection requirement—normally glasses with side impact protection. The bulk of these safety glasses will feature built-in side shields. If you wear glasses to correct vision deficiencies, most shops may allow you to use your regular glasses for safety protection, provided they have shatter-resistant lenses. However, you may have to purchase some sort of side-impact shields to bring them up to the minimum shop requirements. If you wear glass lens glasses, there is a hardening process available that you may want to employ to increase shatter resistance. Wear your safety glasses at all times when you are in the shop area.

Your hands and fingers are also very important. The rules here are simple: keep them away from moving parts. Since gloves are easily caught in moving parts, they should be removed before turning on the machine. Don't even think about reaching for a chip, which is a small, often sharp, piece of metal removed from the workpiece by a cutting tool. It may seem very inviting to reach in and pick chips away from a rotating part, but when you least expect it something else will jump out at you. Your skin makes a poor substitute for a chip raker.

If you must remove small chips from a machine in motion, use a brush. For wiry chips that wind themselves around the workpiece and drill bits, stop the machine to remove them. This doesn't mean to grab them as the spindle slows down. Let the spindle come to a full *and* complete stop.

Many machines have safety shields installed for your protection—not the safety of the machine. Never remove them except for cleaning, and when improved access is required for setting up a job. Then put them back in place immediately before operating your lathe.

Listen…

Learn to use your ears as you work. You can't move the cutter or turn the machine on and off with your ears, but you can learn to distinguish which sounds are definitely out of the ordinary.

A dry-running cutter will produce a slightly different pitch than one that is run with coolant. Dull cutting tools are also easily identified by their peculiar cutting sound. The pitch of the sound made during the machining process can also distinguish fast or slow feed rates.

There will be occasions when you will need to wear hearing protectors—both "ear muff" style protectors and common ear plugs offer protection. Certain materials will create a very high-pitched sound during machining that can be very detrimental to your hearing. In this case, it is not a reflection upon the machinist's ability to select the correct tool; the material is just that difficult to machine. In cases where hearing protection must be worn, be extra careful when listening to shop background noise.

Feel...

By placing your hand on the lathe bed, you will learn how to detect good and bad vibrations. Always feel the machine bed when you start machining your workpiece. If you feel vibrations from the toolholder and not the cross-slide, stop the machining process and find out why. There is a close relationship between your senses of hearing and feeling. This is not to say that you should keep one hand on the lathe and the other on the feed handle. You may want to place your hand onto the bed if you start to hear something out of the ordinary. Or you may want to feel the cutting bit holder when you start the machining of a part to ensure that it is tightened down properly.

Your Responsibility

In the end, <u>you</u> are responsible for your own personal safety in the shop. Shop owners and schools can only provide you with the knowledge and expectations necessary for safe machine operations; it's up to you—the individual—to work safely. Removing existing, or user installed, safety equipment can only lead to system malfunctions and injury to the machine operator. Injuries caused by machine failures are very rare indeed. Tool failures, on the other hand, are common and are almost always caused by the operator. Excessive pushing or crowding of the tool will induce or cause failure. Learning how to compute proper machine cutting speeds and feeds is essential to ensure long and predictable tool life. Exploding tool bits not only cause personal injury, but can be quite costly in terms of tooling and downtime.

Remember, a lathe at rest tends to stay at rest, unless some sort of user interface is applied. The bottom line: don't do stupid things and blame it on the machine. Your safety is most important, so take the time to learn the safety precautions supplied by your lathe manufacturer, and use your own best common sense.

Lathes are heavy. They should be moved and set into place only by experienced people. Do not try to move a lathe by yourself unless you are familiar with safe equipment moving practices.

Accessories used on the lathe may also be heavy. A typical 10" chuck may weigh around 90 pounds (40.9 kilograms) or more. Storing chucks and moving them around the lathe will require some thought. Don't try to lift a heavy chuck all by yourself. Use safe lifting practices. Use your arms and legs when lifting—not your back. Hold the part close to your body and use your arms and legs as the primary lifting force. Holding a heavy object at an arm's length away from your body locates the pressure at angles that will cause back strain. Again, get help when necessary.

The size and weight of certain lathe components or workpieces may require the use of an overhead crane. Be sure to read, understand, and use proper operational procedures before you use any lifting device. See your supervisor or instructor if you are not sure of safe lifting techniques.

When working around certain materials, you may also want to wear a dust protector to keep small particles of debris out of your lungs. Certain materials may also outgas. This is a phenomenon that occurs during machining as the waste material heats up and liberates a gaseous byproduct.

To ensure that you are not working with a hazardous material, consult the appropriate *Material Safety Data Sheet* (MSDS). These sheets are required in some localities, and their contents meet the requirements of the Occupational Safety and Health Administration (OSHA). They contain a wealth of information about what elements make up certain materials, how they react when heated or cooled, what type of firefighting medium is required to extinguish a fire involving them, safety precautions, and other important information that you should pay close attention to. MSDS for many materials can be found on the internet at www.msds.search.com.

Personal Safety Tips

The following tips are for your personal safety

DON'T run your machine until you have read—and understand—all operating instructions.

DON'T run your machine until you have read—and understand—all machine and control key signs.

DON'T run your machine for the first time without a qualified instructor or supervisor to help you when you need it.

PROTECT your eyes. Wear safety glasses with side shields at all times.

DON'T get caught in moving parts. Remove watches, rings, jewelry, neckties, and loose fitting clothes.

PROTECT your head. Wear a safety helmet when working near overhead hazards.

KEEP your hair away from moving parts.

PROTECT your feet. Always wear safety shoes with steel toes and oil resistant soles.

TAKE OFF your gloves before your turn on the machine. They can easily become caught in moving parts.

BEWARE of loose objects—they can become flying particles.

REMOVE all loose items (wrenches, chuck keys, rags, etc.) from the machine before starting.

NEVER operate a machine after taking strong medication, using nonprescription drugs, or consuming alcoholic beverages.

SAFEGUARD the cutting zone (point of operation). Use standard, general purpose safeguards where possible. Use special safeguards when required.

PROTECT your hands. STOP the spindle completely BEFORE changing tools.

PROTECT your hands. STOP the spindle completely BEFORE you load or unload a workpiece.

PROTECT your hands. STOP the spindle completely BEFORE you clear away chips or oil. Use a brush or chip scraper. NEVER use your hands or fingers.

PROTECT your hands. STOP the spindle completely BEFORE you adjust the workpiece, fixture, or coolant nozzle.

PROTECT your hands. STOP the spindle completely BEFORE you take measurements.

PROTECT your hands. STOP the spindle completely BEFORE you open safeguards or covers. Never reach around a safeguard.

PROTECT your hands. STOP the machine BEFORE you change or adjust belts, pulleys, or gears.

PROTECT your hands. Keep hands and arms clear of spindle start switch when changing tools.

PROTECT your eyes and the machine. NEVER use a compressed air hose to remove chips.

KEEP your work area well-lit. Ask for additional light if needed.

DON'T slip. Keep your work area clean and dry. Remove chips, oil, and obstacles.

NEVER lean on your machine. Stand away when the machine is running.

DON'T get trapped. Avoid pinch points caused by motion of the table and head.

PREVENT objects from flying loose. Securely clamp and locate workpiece. Use stop blocks where necessary. Keep clamps clear of cutter path.

PREVENT cutter breakage. Use correct table feed and spindle speed for the job. Reduce feed and speed if you notice unusual noise or vibration.

PREVENT cutter breakage. Rotate spindle in a clockwise (CW) direction for right-hand tools, and counterclockwise (CCW) for left-hand tools. Use the correct tool for the job.

PREVENT workpiece and cutter damage. Never start the machine when the cutter is in contact with the workpiece.

KEEP tools sharp. Dull or damaged tools break easily. Inspect tools and toolholders. Keep tool overhang short.

KEEP rotating cranks and hand wheels well-lubricated and maintained. Do not remove safety guards.

PREVENT fire. Keep flammable liquids and materials away from work area and hot chips.

PREVENT fire. Certain materials, such as magnesium, are highly flammable in dust and chip form. See your instructor or supervisor before working on such materials.

PREVENT the machine carriage from moving unexpectedly. Disengage power feed when it is not being used.

PREVENT the machine from moving unexpectedly. Always start the machine in manual mode.

Courtesy, Bridgeport Machine Tool

Recap and Review

There is no end to safety in the machine shop, and the same is true in this book. You will find safety precautions and suggestions throughout the book. Pay attention to all of them, they are not here just to fill the book with more pages. Your personal safety is paramount to everyone in this business, and most of all to yourself.

Review Questions for Chapter 1

1. Glasses with _____ offer better protection than those not equipped with them.
   A. Ear protectors
   B. Neck cords
   C. Wire frames
   D. None of these.

2. The best jewelry to wear while operating a lathe is . . .
   A. Gold
   B. Silver
   C. Platinum
   D. None at all.

3. Machine safety shields are installed for _____ protection.
   A. Your
   B. The machine's
   C. No protection, and should be removed
   D. None of these.

4. Who, in the shop, is responsible for your personal safety?
   A. The machine manufacturer
   B. The owner of the shop/school
   C. Your supervisor
   D. None of these.

5. Which of the following should you use when lifting heavy objects?
   A. Your legs and back
   B. Your arms and legs
   C. Your back and arms
   D. None of these.

6. What is OSHA?
   A. The occupational, safety, and housing administration
   B. The place in Wisconsin where they make the cheese
   C. The occupational salary and housing authority
   D. The occupational safety and health administration.

7. PROTECT your hands: _____ the spindle completely before changing tools.
   A. Stop
   B. Start
   C. Slow
   D. None of these.

8. Gloves should be _____ before you start the lathe.
   A. Put on
   B. Adjusted for fit
   C. Removed
   D. None of these.

9. To prevent cutter breakage, _____.
   A. Rotate the spindle counterclockwise (CCW)
   B. Rotate the spindle clockwise (CW)
   C. Increase feeds and speeds
   D. None of these.

CHAPTER 2
Machine Maintenance

No matter what machine you operate in the shop, maintenance is vital. The lathe is no exception. Without routine and daily maintenance, your lathe will be an expensive heap. Your lathe depends upon lubrication—the lifeblood of any machine.

Lathe manufacturers can and will provide you with a chart of recommended daily, weekly, monthly, and yearly maintenance. If you stick to this schedule, your machine will return many years of trouble-free operation. It is not uncommon to find lathes 40, 50, 60 years, and older, that are still capable of producing high quality work because they have been maintained properly.

Lubricating Your Lathe

The basics of machine maintenance include daily application of *way* oil to those machined surfaces that are openly exposed, e.g., ways, cross-slides, leadscrews, etc. Ways are the highly finished areas on the bed on which the tailstock and carriage ride, and they are not protected by paint or other coatings. Way oil is specially formulated to provide not only lubrication, but also a certain amount of stickiness to keep it in place after it is applied. This is a good attribute because we don't want the machined surfaces to rust. Rust is the enemy of all machines. The corrosiveness of iron oxidation (rust) will eventually cause the machine to become inaccurate. When applying way oil to the machine, don't skimp; apply a liberal amount of oil to the machined surfaces and leadscrew.

Way oil comes in three basic weights: *light*, *medium*, and *heavy*. For simplicity, purchase the medium grade, which is required by the majority of lubrication points. Again, your lathe manufacturer can specify the exact weight (light, medium, or heavy) for each component on the machine. You may be surprised to find that the weight of the oil that is recommended may vary from one workpiece material to another.

Figure 2.1 *Felt wipers are an integral part of the maintenance routine; without them, small chips and debris can become lodged beneath the carriage and cause accelerated wear to the machine ways.*

An integral part of the lubrication process will be to maintain the wipers made of felt material that are attached to the carriage. These wipers are used to lubricate and clean the ways when the carriage is moved from side to side. The felt may require periodic replacement as it becomes worn and loaded with minute chips and debris created during the cutting process. Be careful not to hit the felt wipers with parts or tools. They are somewhat fragile and can be damaged without notice. The cross-slide may or may not have felt wipers attached—the larger the lathe, the more likely it is to have wipers.

There are a number of different methods used to admit oil into the nonvisible bearings, gears, and sliding surfaces on the lathe. The most popular are the ball-detent type because they can be installed without impeding the operation of the lathe, and will not get caught on clothing, etc. The ball must be depressed in order to allow the oil to pass into the orifice. Some of these ball-detent types may require the use of a small screwdriver to hold the ball open to allow you to pour lubricant into the lube orifice. A cup-type oiler is commonly used on the leadscrew, and in other areas where more than just a few drops of lube are required. The cups have a self-closing lid to keep out contaminants once the oil is poured in. Always fill the cup to the top of its reservoir. The use of *oil drippers* has decreased since the installation of precision ball and roller bearings started to make their appearance on lathes many years ago. The oil drippers were pretty nifty. You filled them up with oil, adjusted a small valve to regulate the number of drips per minute, and went to work. When the reservoir was empty, you filled it up again, and went on. You could always look at the glass reservoir and see exactly how much oil was left.

Some lathes will require a certain amount of grease lubrication. This type of application is always best performed with a hand operated grease gun: load a cartridge of the specific type of grease into the gun; then slip the gun socket onto the grease nipple (properly called a zerk fitting), and give two or three quick squeezes. Don't overdo it because overgreasing can impair machine operation, or create a messy work environment.

For those maintenance and service items that require the disassembly of the spindle bearings, call in someone who is familiar and qualified for this type of service. Disassembly of the lathe without regard for the exact fit of certain components will cause your lathe to become inaccurate and make it impossible for you to produce high quality parts.

Cleaning Your Work Area

The area around the lathe should also be maintained with a certain degree of diligence. If you let chips pile up, you will most likely not be working in a safe area. Your instructor and/or boss may not appreciate your lack of attention to these details. Machine and shop cleanup is your responsibility, not someone else's. Spend a few minutes (a couple of times a day) to keep the machine shipshape. There are some materials that make quite a mess when you machine them. If you start to get big piles of chips, stop what you are doing and clean them up. Don't wait for them to become a hazard and cause you harm.

Once a week, you should clean your machine with an approved degreaser. Increasingly, companies are converting to water-based cleaning solutions. These cleaners are mixed at a specific ratio, or concentration, of cleaner to water (check your cleaning bottle label for specific directions). Application of these cleaners is typically performed with a spray bottle. Spray a liberal amount of the solution onto the painted surfaces of the machine, but not onto the machined surfaces! Allow the solution to break down the grease and grime. Once the grime starts to roll off the surface, take clean cloths or paper towels to wipe the surface to remove the remaining debris. Dispose of

Chapter 2 - Machine Maintenance   17

the towels in an approved manner to prevent their eventual disposal into a sanitary landfill (very bad for the environment). Notice that we didn't apply the solution to the machined surfaces; this is because the water base may

NEVER use flammable liquids such as gasoline, kerosene, or acetone to clean machinery!

cause rust. If the ways of the machine do require cleaning, wipe the affected areas with a cloth or paper towel to remove the mess.

Make sure that you keep the floor clean and free of oil and other lubricants. It's amazing just how many people slip and fall and cause themselves great bodily harm! Taking a few minutes to clean up a spill can save a whole pile of money and time off from work. Keep it clean!

Using Cutting Fluids Properly

Application of the cutting fluids (coolant) is one of the major tasks that you will be responsible for when operating a lathe. By and large, the

Figure 2.2 Flood cooling can create a cleaning problem if the fluid is allowed to be thrown off the workpiece during machining.

messes created by lubricants will exceed those of the machined chips. More information is presented later in the book, but let's preview some of the maintenance tasks. Coolant application will fall into one of two primary categories: flood and applied.

Flood coolant application is when a constant stream of coolant is applied to the surface of the workpiece during the machining process. Care must be taken to control the *throw-off* or splashing of the coolant. Coolant that is thrown off by the spinning workpiece can create huge cleanup problems, and may contribute to environmental problems.

Coolant can also be applied by misting, spraying, or brushing it on. The amount of coolant applied with these methods normally results in less mess.

To clean up excess coolant, use the aforementioned spray cleaner that we discussed above. Your lathe may provide various types of filter media

Figure 2.3 *The use of a brush to apply coolants and lubricants will decrease the amount of fluids used, and lessen the amount of cleanup time.*

and magnetic separators to clean the chips and debris (commonly called swarf) created during the machining process from the coolant, so clean these as required. Be sure that all coolant levels are properly maintained because this will ensure long and reliable pump life.

Other Scheduled Maintenance

Once every six months, in order to ensure proper cutting action, the lathe should be checked to determine that it is level. Failure to check this critical maintenance item may cause long workpieces to become tapered during the machining process. Taper is a uniform and gradual decrease in size, resulting in a cone shape. It can be checked by placing a long piece of stock (the longer the better) between the lathe centers. A diameter is then turned on the end nearest the chuck or faceplate, then the same diameter is turned at the opposite end of the stock. **Note:** It may be necessary to center-drill the ends of the stock in order to turn the piece correctly. Both cut diameters should be the same, or within a few tenths, to indicate that the lathe is leveled properly. This process will be discussed in more detail later in the book.

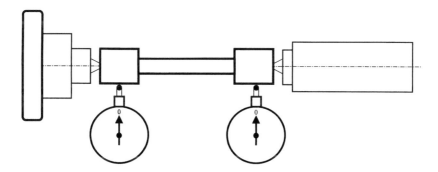

Figure 2.4 The use of a test bar is a quick and easy way to verify that your lathe is perfectly level.

Later in the book you will also learn about *backlash*. Essentially, it is the slack or wear between two rubbing surfaces, as would be the case with the cross-slide and compound-rest screws and nuts, and the dovetail slides, which are made up of *gibs and ways* (gibs are pieces of metal, often wedge shaped, that locate machine components on the way). The *half-nut* adjustment for the leadscrew should also be made, and the procedures for maintaining each of the adjustments on your lathe will be contained in the lathe operation handbook. If your lathe does not have an operation handbook, or if you are unsure about how to make these adjustments, employ the services of a professional machine mechanic to perform this maintenance.

Troubleshooting Chart

Trouble	Problem Cause	Correction
Vibration	Loose leveling screws	Set all screws so they are even on floor
	Torn or mismatched belts	Renew v-belts with matched set
	Work or chuck out of balance operating at high spindle speeds	Balance chuck or reduce spindle speed
	Motor out of balance	Contact motor manufacturer
Chatter	Cutter bit improperly ground or too wide on area of contact	Regrind cutter bit or adjust toolholder
	Tool overhang too great	Keep point of cutter bit as close as possible to centerline of tool post/holder
	RPM too high	Reduce RPM
	Feed rate too high	Reduce feed rate

(Continued on next page)

Troubleshooting Chart *(continued)*

Trouble	Problem Cause	Correction
Chatter	Gibs of cross-slide or compound-rest loose	Adjust gibs
	Work improperly supported	Adjust tail stock center, use steady rest or follower rest on long slender shafts
	Vibration	See "Vibration" above
	Spindle bearing loose	Adjust or replace spindle bearing
Work Not Turned Straight	Head stock and tail stock centers not aligned	Align tail stock center
	Work improperly supported	Use steady or follower rest. Reduce workpiece overhang from chuck
	Bed not level	Relevel bed using precision spirit level
	Tool not on-center when using taper attachment	Put tool on-center
Work Out of Round	Work loose between centers or centers are excessively worn—work centers out of round	Adjust tail stock center. Regrind centers, lap work centers
Cross-slide or compound-rest movement does not coincide with dial movement on respective adjusting screw	Gib setting too tight or too loose	Adjust gibs
	Work is too long and slender	Use steady or follower rest

Machine Maintenance

Daily	Weekly	Monthly
Clean & lube ways	Inspect for wear or damage	Inspect for wear or damage
Lube spindle bearings	Inspect gibs	Inspect belts
Lube lead & feed screws	Clean machine thoroughly	Lube drive gears
Lube tail stock & quill		Inspect spindle bearings
Inspect for wear		

Review Questions for Chapter 2

1. _____ oil should be used to keep the machine ways lubricated and to protect them from rust.

    A. Motor

    B. Machine

    C. Way

    D. None of these.

2. Felt wipers help to _____.

    A. Spread the oil evenly

    B. Keep chips out of the ways

    C. Keep chips out of the head stock

    D. Keep chips out of the tail stock

3. Lubricating oil should be used _____.

   A. Liberally

   B. Sparingly

   C. Not at all

   D. None of these.

4. You are responsible for cleaning your work area.

   A. True

   B. False

5. How often should you degrease your machine?

   A. Every day

   B. Once a week

   C. Once a month

   D. Never.

6. *Throw-off* is associated with the use of _____ _____.

   A. Speeds & feeds

   B. Cutting pressure

   C. Cutting fluids

   D. Thrust angle of the cutting tool.

7. The lathe should be checked for level every _____ months.
   A. 1
   B. 3
   C. 6
   D. 12.

8. A test bar is used to _____.
   A. Ensure that the lathe bed is level
   B. Check alignment between the head and tail stocks
   C. Check alignment between the head stock and the cross-slide
   D. None of these.

9. All of the following things should be checked on the lathe weekly, except:
   A. Lube drive gears
   B. Inspect for wear or damage
   C. Inspect gibs
   D. Clean machine thoroughly.

Chapter 3

Engine Lathes

In this chapter you will start to develop a sense of what it takes to operate a lathe. As you start your journey, develop a keen curiosity about how and why the lathe works, and what it takes to operate each component of the machine. Learn how to be a machinist, not just an operator. As a machinist, you will be required to make many decisions that affect the successful outcome of transforming a piece of raw stock into a finished part.

Some History

The earliest form of turning machines involved placing a log between two trees, with one end of a rope tied to a branch and the other to the log. Winding the rope around the log pulled the branch taut. When the log was released the branch would turn the log, at which time a cutting tool would cut a shape into the workpiece. It was pretty simple, but also pretty slow. At a later date, when machines moved indoors, a strip of wood or *lath* would be attached to the support the rope. This may possibly be the reason why we use the term *lathe* to describe a turning machine.

The first mention of a *tourneur* was recorded in France around 1740. A diagram of the device was published in 1741. It used a hand crank to turn a workpiece held between two centers. A leadscrew used to move the cutting tool was geared to the hand crank, but no provision was made for changing the cutting feed rates.

Enter Henry Maudslay, an Englishman who developed the first practical lathe used to cut screws. The lathe was fashioned with gears that could be changed for different thread sizes. These interchangeable gears connected the spindle and leadscrew.

The first lathes built in the United States were constructed with wooden beds (essentially a tree trunk that was sawn into a square log) equipped with iron ways. The fitting of iron ways greatly increased the serviceability and accuracy of the lathe. Around 1836, Putnam of Fitchburg, Massachusetts, built a small lathe with a leadscrew. In 1850, lathes made of iron were manufactured in New Haven, Connecticut. A 20" swing by 12' bed lathe with a back-geared head was produced in New York. ("Swing" is determined by the distance from the center of a lathe's spindle to the nearest point on its ways—the term is used to denote the largest diameter workpiece that the machine will accommodate.) All of these contributions played a large role in the American Industrial Revolution. Without these innovations, we might all still be riding around in horse drawn buggies and using sailing ships to cross the great oceans of the world.

Lathe Types

The engine lathe is a basic staple in any machine shop. It is the only machine in the shop that can (if large enough) reproduce itself. The machine has a limited amount of actual operations that it can perform, but when these combinations are sequenced, the lathe becomes a very powerful tool indeed. The lathe shown in *Figure 3.1* is a conventional engine lathe. The controls and accessories on this lathe may not be exactly like the lathe you are using, but they will be similar. The functions of the fundamental components are as follows.

Motor Drive and Head Stock: Transmit power to rotate the workpiece.

Head Stock and Tail Stock: Hold or support the workpiece.

Head Stock and Electrical System Controls: Provide control and various speeds of work rotation.

Tool Post and Compound-rest: Hold the cutting tool in position.

Carriage, Apron, Quick Change Gearbox, and Leadscrew: Provide hand and power movements (at various rates) of the cutting tool.

Bed and Legs: Provide steady framework for all other components.

Leadscrew and Chasing Dial: Provide controls for threading operations.

Chip Pan: Provides collection of chips or cutting fluid.

Basic engine lathe operation has not changed significantly since the Industrial Revolution: hold the workpiece, turn it, and shape it—simple. The size and capacity of the lathe has, however, changed many times. Some lathes are so small (often referred to as jewelers' lathes) that they can fit very neatly onto a small workbench, while others may require an entire building to house them.

A lathe is classified or sized by the maximum diameter and length of the workpiece it can safely handle. Notice the word *safely*. It may be

Lathe Operation and Maintenance

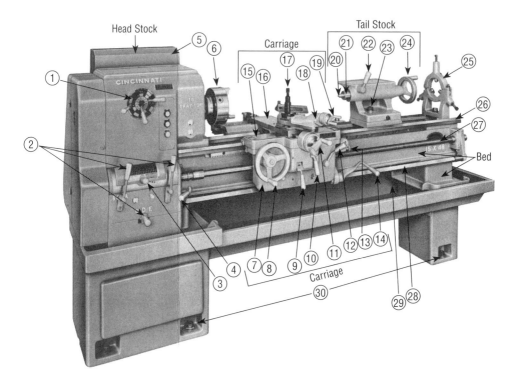

Figure 3.1 A conventional engine lathe. (Cincinnati Lathe and Tool Co.) The components are as follows. #1. Spindle speed control selector. #2. Threading and feed selection levers. #3. Quick-change gearbox. #4. Spindle Start/Stop lever. #5. Electrical system control box. #6. Spindle with chuck. #7. Apron. #8. Carriage handwheel. #9. Power feed engaging lever. #10. Cross-slide power feed engaging lever. #11. Cross-slide adjusting handle. #12. Half-nut engaging lever. #13. Thread chasing dial. #14. Spindle Start/Stop lever. #15. Carriage saddle. #16. Cross-slide. #17. Tool post. #18. Compound rest. #19. Compound-slide adjustment handle. #20. Dead center. #21. Sleeve, or ram. #22. Tail stock spindle clamp. #23. Tail stock clamp nut. #24. Tail stock spindle adjustment wheel. #25. Steady rest. #26. Bed ways. #27. Leadscrew. #28. Feed rod. #29. Chip pan. #30. Leveling jacks.

possible to modify or exceed the safe maximum working limits of a lathe, but it is not wise to do so. If a lathe is classified as 12" by 36", the machine can safely accommodate a center-mounted workpiece having a maximum diameter of 12" (its "swing") and a maximum length of 36". Some lathes, known as *gap-bed* lathes (*Figure 3.2*), have a removable section of bed that allows the mounting of large diameter workpieces. This is a handy feature when turning automotive flywheels, large diameter pipe flanges, or similar workpieces.

Through the years, the lathe has been modified to increase production output. One way this has been accomplished is by the addition of a special turret-indexer placed onto the lathe bed. The turret lathe, as it is called, allows for the installation of several different tools on the turret. Each time the turret lever is activated, the turret will spin and select (or "index") the

Figure 3.2 *An example of a gap-bed lathe.*

next tool in sequence—it is then ready to carry out a different machining operation. This means that the machinist does not have to manually change tools each and every time the machining operation requires a different tool.

An additional feature found on many modern lathes is a *depth control* attachment that permits the machinist to simply preset the machining depth and then move the tool into and out of the workpiece until the desired depth is reached. Without a depth control attachment, the machinist has to depend upon a manual gage reading each and every time the operation is performed. Some turret lathes also feature a *push-pull* cross-slide (the cross-slide is the device that feeds the cutting tool into the workpiece) that can be moved transversely (in and out) with a lever. On other lathes, a feed wheel must be turned to move the cross-slide.

Computer numerical controls are now used on most top-of-the-line lathes. On a CNC lathe, the controls that are normally operated by a machinist are instead activated by a computer program that, in effect, becomes the machinist, although a machinist is still necessary to set up the lathe and write the required program. If the controller fails, production stops dead in its tracks, but when the controller and tools are all set up, even the best machinist is no match for the number of completed parts that can be produced in an hour, nor can a machinist make them as accurately.

Another variety of the lathe is called the *speed lathe*, or manual (since it is not equipped with automatic feeds) chucker. This machine was designed first and foremost with production in mind. It has a turret arrangement where the tail stock would normally be located, which means that the speed lathe cannot be used to machine long parts. The cross-slide has a large diameter dial that allows the operator to machine parts within a few ten-thousandths of an inch of each other. A large dial is used, because the larger the dial, the larger the space between the numbers; larger spacing between the numbers allows for more accurate work. Speed lathes are also fitted with a *collet closer* rather than using a *chuck* to hold the workpiece. Collets and chucks are clamping devices used to securely hold a workpiece on lathes and other machine tools, and they will be discussed later in the book.

Lathe Construction

Lathes are made to last. The main components of the lathe are always made from cast iron, which, for our purposes, has several advantages over steel. First of all, it can be made relatively inexpensively as compared to steel. Second, and one of its more important merits, cast iron machine bases are stable and do little to transmit unwanted vibrations as the machine runs. This is a very important point. When a cutting tool is engaged into a workpiece, oscillations are set in motion that, if uncontrolled, can leave behind an undesirable surface finish on the workpiece. Cast iron acts like a sponge when it comes to vibrations—it soaks them up very effectively.

Another reason for using cast iron for lathe construction is that the amount of thermal expansion experienced during operation is minimal. When any machine is used or operated, it generates heat. Heat, as we all know, causes things to expand. If a machine is constantly expanding and contracting, these changes in size will impact the dimensions of the parts being machined. This may seem like a small point right now, but if you were to produce a part from an expensive piece of material, and the finished part failed to meet dimensional requirements due to machine expansion, you would quickly gain an appreciation of cast iron's relatively favorable thermal characteristics.

The structure of cast iron allows a controlled amount of oil to penetrate the surface and stay resident, which contributes to the lathe's lubrication. In most circumstances, a minimum film of oil (.0001", or 0.0025 mm) should be present on all contacting metal surfaces. When metal parts are allowed to rub against each other without lubrication, significant amounts of wear will very rapidly become apparent. Add a small amount of abrasive material like steel chips and grit, and the wear increases. As you read in the chapter on maintenance, felt wipers plus liberal lubrication are essential to prevent the migration of abrasive bits and pieces from wearing out the machined surfaces.

Next, we will take a close look at the individual components of the engine lathe. We will start at the front (or left side as the operator faces the lathe) and move to the back while taking a close look at every major component we encounter. You will probably want to refer to *Figure 3.1* to locate and identify specific components and controls.

Head Stock

There are some basic controls used to operate an engine lathe, and the most fundamental are located on the front or top of the head stock. Each machine control is typically operated with a handle, lever, dial, or switch. Locate and learn how to operate the power on-off switch and spindle speed controls before performing any type of machining operation on your lathe.

Local or national governmental organizations may require a separate wall-mounted electrical control box to enable *fail safe* shutdown for the lathe in addition to the machine mounted controls.

Voltages required for lathe operation will vary from country to country, as will the electrical frequency (Hz). Lathes manufactured for use in the United States and other countries will use 60 Hz electricity, while some countries use 50 Hz current. The actual voltage will depend upon your electrical service to the building. For *home shops*, 110 VAC 1ϕ (single phase), or 220 VAC 1φ, will serve very well. Single-phase electrical circuits are typified by three (3) spades on the electrical connector: two (2) conductors and one (1) ground.

Industrial buildings will be serviced with 220, 440, and 660 VAC 3φ (three phase) electricity. Three phase electricity is more efficient than single phase. Their identifying characteristics are plug ends with four (4) prongs: three (3) conductors, and one (1) ground.

> NEVER attempt to service your electrical circuits unless you have completed and passed proper training in this field. Qualified electrical technicians should always be called when electrical problems arise. You may be electrocuted if you don't know what you are doing. Read, follow, and understand all directions for machine safety before you operate your machine or power tool.

On-Off switch

The first control you need to locate and become familiar with is the On–Off switch or lever. Depending upon the age and manufacturer of your lathe, the machine may be equipped with a simple *drum* switch or *clutch* lever to run the spindle *forward* and *reverse*. Some machines may have both: a drum switch to start the motor, *and* a clutch lever to start and stop the spindle. The drum switch is easy to operate—simply turn the switch from the off position to either the forward or reverse position. Be sure that you keep your hands and body away from the spindle whenever you turn the machine on or off.

On machines fitted with clutch levers, you will need to determine whether the handle must be *pulled up* or *pushed down* to rotate the spindle forward and in reverse. Some clutch lever-type controls also have a built-in *brake* feature to rapidly slow the spindle to a stop. If this is the case, you will need to determine if the machine has a *five-way* control lever: the positions are (1) forward-stop, (2) forward-on, (3) neutral, (4) reverse, and (5) reverse-stop. If you have a choice of what direction the machine runs when it is initially hooked up to the electric service, it should be wired as follows. Have the drum switch wired so that when the switch handle is turned *right*, the spindle turns clockwise (CW), and when the handle is turned *left*, the spindle turns counterclockwise (CCW). Have the clutch lever wired so that when it is pushed *down* the spindle rotates clockwise, and when pulled *up* the spindle rotates counterclockwise. The spindle **must** be brought to a full and complete stop before changing spindle direction; otherwise serious damage to you or the machine may result. Be careful!

Spindle speed controls

The spindle speed controls are mounted on the operator's side of the head stock. Usually, they consist of two or more handles that are clearly

Figure 3.3 *The lever-type handles are very easy to use. Some lathes have a built-in brake feature to stop the spindle quickly.*

marked, so you should not have any difficulty understanding their functions. Depending upon the age, manufacturer, and type of lathe, the speed controls will vary.

The mechanical components of the drive system are enclosed inside the head stock, and there are two basic types of spindle drive systems—*belt* and *gear* drives. As the name implies, belt drive machines use one or a series of V-belts to drive the spindle. This system is further classified as *manual-speed change* or *variable-speed change*. The manual speed change type machines move the drive belts from one pulley to another to change spindle speed. Moving the belt from a small diameter pulley to a larger pulley will, in most instances, increase spindle speed. The machine must come to a complete STOP when making speed changes from one pulley to another. The downside to manual-speed change machines is the limited number of speeds available. Having only 8 or 12 speeds may not fit all applications.

In most cases, machines equipped with variable speed controllers must be running to make speed adjustments. Failure to run the spindle during speed change may cause damage to the lathe, and possibly the operator. Always read your factory supplied lathe operator's handbook before operating the machine. It is essential that you follow all the safety rules and precautions.

Gear driven lathes also have belts that connect the motor to the power input shaft on the lathe. However, these belts do not have to be changed or adjusted to affect spindle speed or rotation. To change spindle speeds on gear driven lathes, you need to ensure that the motor is turned completely off and that the spindle rotation has come to a stop. This is because the lathe's transmission gears are not synchronized, like an automobile's transmission, and the engaging gears will crash with one another if the spindle is rotating.

Remember, spindle speed change is effected when levers and/or dials are changed from one position to another. The levers/dials should be clearly marked so that there is no mistaking where they should be positioned to match a particular speed.

Some lathes, specifically those with belt drives, will be fitted with a device known as a *back-gear*. A back-gear allows the spindle to turn at very slow speeds. This means that a four- or six-speed machine will actually be capable of eight or twelve speeds. You will find that some lathe operations, like threading, <u>must</u> be performed at low spindle speeds. To actuate the back-gear, the lathe must be turned off. There will be two levers or knobs used to control the back-gear operation. Check with your factory machine operator's handbook for exact procedures for engaging the machine into and out of back-gear.

Spindle

The *spindle* is the business end of the lathe. Part holding devices are attached to it, and it turns the workpiece. Without the spindle the lathe would be pretty useless. The spindle's drive is housed in the head stock where gears or belts are used to provide the motive force. All modern spindles are supported with rolling element bearings. Older lathes cradled the spindle with a bronze or brass sleeve-type bearing, which worked fine until they began to show signs of wear. Then the spindle would move off-center and cause the workpiece to become tapered. Newer lathes use massive ball or tapered roller bearings to support the spindle.

There are four basic types of spindle noses, or face mounting plates: threaded (*Figure 3.4*), short taper (*Figure 3.5*), cam lock (*Figure 3.6*), and long taper key drive type (*Figure 3.7*). Each one does the same job—it mounts the chuck or faceplate to the spindle—but each employs a different means. Spindle styles and face mounting plates are described in detail in Chapter 6.

One of the considerations of the lathe buyer is the *spindle bore diameter*. The spindle bore must be large enough for stock to be passed through it. This may seem like an insignificant point, but when machining a continuous rod, the larger the stock that will pass through the spindle, the larger the workpiece the machine will be able to accommodate.

Figure 3.4 An example of a threaded spindle. (Courtesy, Toolmex Corporation.)

Figure 3.5 An example of an A-type short taper spindle. (Courtesy, Toolmex Corporation.)

38 Lathe Operation and Maintenance

Figure 3.6 An example of a D-type camlock spindle. (Courtesy, Toolmex Corporation.)

Figure 3.7 An example of an L-type long taper key type spindle. (Courtesy, Toolmex Corporation.)

Lathe Bed

The *bed* is the heaviest part of the lathe. On most lathes, it is constructed of cast iron, which is very stable under a wide range of conditions. Cast iron holds its shape in hot and cold weather, and it does not *ring* like steel (steel has a crystalline structure) when it is struck by a hammer or other metal object—instead it tends to absorb noises and vibrations.

The top surface of the bed is precision machined. This provides a smooth sliding surface for the carriage saddle to ride on. A pair of rails, known as *ways*, are located on the top of the bed and will be either flat or "V" shaped. On many lathes the ways will be machined directly from the cast iron bed material, but later designs use hardened and ground steel, which is much more durable. The ways guide the carriage and tail stock, and they are used to align the head stock to the bed. Care must be taken not to damage the surface of the ways.

Depending upon the construction of the lathe, the machine legs or pedestals may be cast integrally or separately from the bed.

Quick-Change Gearbox

The quick-change gearbox is mounted onto the front of the bed. It engages the leadscrew or feed rod to drive the carriage for turning and threading operations. The leadscrew is a large threaded rod that is located below and parallel to the ways, and on most lathes is used for cutting screw threads. On less sophisticated lathes, the leadscrew will be the only drive mechanism available, and will be engaged for all cutting operations. However, because the leadscrew is a precision mechanism and can be prone to damage, better lathes have a primary drive mechanism called the feed (or control) rod that engages gears in the apron to move the carriage assembly longitudinally for all operations other than threading. For turning operations, the quick-change gearbox automatic feed control should be in the "on" position to allow it to use the feed rod to advance the carriage. For threading, the half-nut (sometimes called the split-nut) lever on the apron is engaged, and the leadscrew will then provide the drive to advance the carriage.

At first glance, the quick-change box may look a little confusing. To help explain its controls, there is usually a chart affixed to the head stock (see *Figure 3.8*) that contains rows and columns of pairs of numbers that provide options you can select for turning and threading operations. One of the selections will be the tool feed rate, expressed in decimals of an inch (or millimeters) per revolution. An example of feed rate would be .0032"

Figure 3.8 This is an example of what a quick-change gearbox chart might look like. Note that this chart may be different from the one on your lathe. Check your lathe's operator's manual to get exact specifications.

per spindle revolution, or 0.08 mm per spindle revolution, which means that every time the spindle completes a 360° rotation/revolution, the carriage advances .0032", or 0.08 mm.

In addition, there are usually two handles on the gearbox that you must understand before you can put the lathe in motion. Once you have deciphered the chart, and the function of the handles, you will be able to set the gearbox to move the carriage unit a predetermined distance during each spindle revolution.

Because the chart details and chart–handle relationship differs with almost every type and brand of lathe, it is essential that the operator's manual, or a qualified instructor, be consulted before using the quick-change gearbox on your lathe.

Carriage

While the principal job of the spindle is to rotate the workpiece, the carriage performs two functions: 1) it moves the cutting tool longitudinally (parallel to the workpiece), at a rate determined by the operator and set at the gearbox, and 2) it moves the cutting tool transversely (perpendicular to the workpiece) to remove material from the workpiece to a preset cutting depth. The complete carriage assembly is shown in *Figure 3.9*. The carriage contains three principal components: the saddle, the apron, and cross-slide which incorporates the compound-rest and tool post. The saddle is essentially a cast iron frame, shaped like the letter "H," that rides on the machine ways and locates and houses the cross-slide and apron. Other parts of the carriage are discussed below.

Apron

Located on the front of the carriage is the apron, which contains the gears and shafts that control carriage movement. Positioned on the apron are the carriage handwheel for manual movement of the carriage assembly; the cross-slide (transverse) feed wheel; the half-nut (or *split-nut*) control lever that engages longitudinal feed for threading; and the threading (or *chasing*) dial or handle. These are the basic controls included on almost every lathe, but most modern lathes will also have a power-feed control (in addition to the cross-slide feed wheel) to move the cross-slide across the top of the carriage.

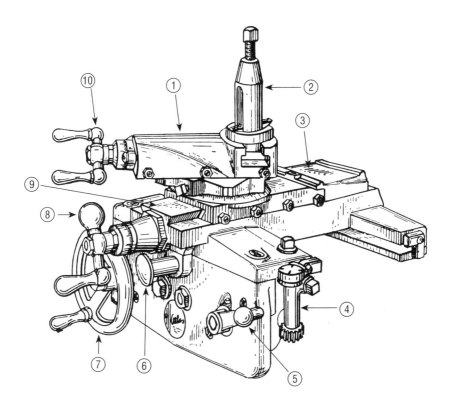

Figure 3.9 Details of the carriage assembly. #1. Compound rest. #2. Tool post. #3. Cross-slide. #4. Thread chasing dial. #5. Half-nut engaging lever. #6. Power cross feed engagement. #7. Carriage handwheel. #8. Cross feed dial and handwheel. #9. Dovetail slide. #10. Compound slide adjustment handle.

The half-nut closure lever was mentioned earlier when discussing the quick-change gearbox. The lever is generally attached to the side or on the front of the apron. Unless you are constantly using the lathe for threading, the half-nut control will be disengaged from the leadscrew to preserve its accuracy and prevent premature wear. When engaged, the half-nut should not be tightened excessively against the leadscrew—not too tight, not too loose.

The threading dial or handle is used in threading operations and lets the operator know when to close the half-nut. It precisely sets (or "*indexes*") the leadscrew and carriage positions in relationship to the workpiece, and it allows the lathe to repeatedly return to the exact same position. This is especially important when cutting threads because if the leadscrew is engaged too soon or too late, the threads will be double cut. In other words, you will wipe out the threads that you just cut.

It may be necessary to place the longitudinal and transverse feed handles into neutral position before the threading dial can be engaged.

Cross-Slide

The *cross-slide* is located directly on top of the saddle and is placed exactly 90° to the lathe bed orientation. The primary purpose of the cross-slide is to move the cutting tool transversely into the workpiece.

The cross-slide uses a *dovetail way* for alignment. An adjustable *gib* is installed on the way to provide for wear compensation. To protect the feed screw from becoming fouled with chips, a cover or plate is typically attached to the top of the cross-slide. The cover or plate moves in and out with the cross-slide. The cross-slide is moved with a feed wheel/handle or by power assist. A graduated scale on the feed handle is used to indicate linear movements.

Compound-Rest

Positioned on top of the cross-slide is the compound-rest (or compound-slide). The compound is similar in construction to the cross-slide—it also uses dovetail ways for alignment, and a feed handle is used to control movement and travel. Like the cross-slide, a graduated scale on the feed handle is used to measure tool movement.

The compound-rest provides the base for the tool post, which contains a slot for positioning the toolholder and cutting tool.

The important feature of the compound is that it can be rotated for positioning the cutting tool at angles other than 90°. This means that cuts onto the surface of the workpiece can be made at angles other than 90°. This is a very handy feature when machining chamfers or when cutting screw threads.

Tail Stock

The tail (or foot) stock is a movable part of the lathe that sits atop the

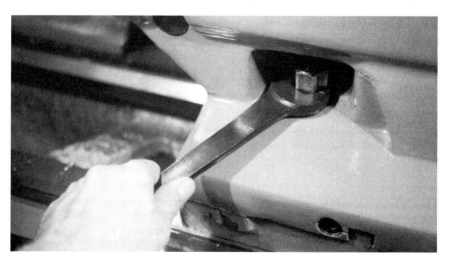

Figure 3.10 The tail stock is locked in place by tightening a handle or wrench.

bed and opposite the head stock. The tail stock has one or more grooves cut into its bottom—they rest on the ways and align it with the head stock. The tail stock can be clamped in place with a wrench or lever that extends or retracts a clamp bar. Do not over- or undertighten the clamp bar because excessive tightening can cause wear on the machine or cause the clamp bar to bend or break. Undertightening will allow the tail stock to move during machining operations. This may cause a dangerous situation whereby you and/or the machine may become injured or damaged.

The primary control on the tail stock is the feed wheel. Turning the wheel clockwise causes the tail stock sleeve (or *ram*, also sometimes referred to as the tail stock spindle) to move toward the head stock; turning it counterclockwise retracts the sleeve. The tail stock sleeve incorporates a very simple ruler to indicate in and out movement. These scales are typically calibrated in $^1/_{16}$" and/or 1.00 millimeter increments. Obviously, this is not a precision control, and it should be used for general measurements only.

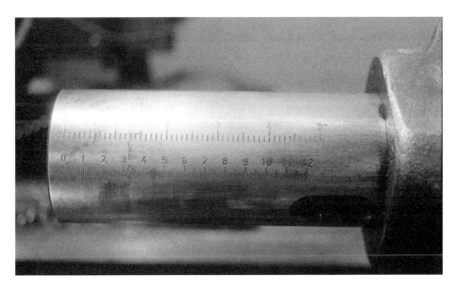

Figure 3.11 *The tail stock's sleeve has a graduated rule that can be used to measure stroke length.*

One of the features that makes the tail stock very handy is that it can be positioned *offset* to the spindle. This means that the centerline of the tail stock can be moved from side to side. When a workpiece is cut with the tail stock offset in the direction of the cutting tool (toward the operator), a taper is cut onto the part: the end of the workpiece located near the tail stock will be smaller in diameter than the end near the head stock. By changing the offset by moving the center of the tail stock away from the cutting tool, the small end will be located near the head stock. **Note:** When offsetting the tail stock, you will have to use a lathe center, face plate, and drive dog on the workpiece, otherwise the taper will not be cut exactly as you like. There will be more on this later.

The tail stock has a hole machined into the sleeve to allow for the installation of tools. The hole at first appears to be round and straight, but upon closer inspection you will find that it is slightly tapered. The taper used in the tail stock is called a *Morse taper*, and it is available in several sizes. The majority of lathes, save very small jewelers' types, use Morse tapers ranging in size from #1, #2, #3, #4, #5, to #6. The size of the Morse taper on your tail stock was determined by the factory when it was first assembled. Small lathes with 6" to 9" swings may use a #2 Morse taper, machines with 9" to 16" swings may have a #3. Generally, the larger the lathe, the larger the Morse taper in the tail stock.

In addition to holding a center for supporting the free end of the workpiece, several tools can be used in the tail stock sleeve including drills, taps, reamers, and boring tools. This allows for a variety of machining operations to be performed.

Tools are easily removed from the sleeve by turning the feed handle counterclockwise (CCW) until the tool pops out. **Note:** Be sure to hold onto the tool while you are removing it, as failure to do so may cause irreparable damage to the tool and/or the machine ways and bed.

A series of Morse taper adapter sleeves are available to allow the use of different size Morse taper tools in the tail stock. More information about these adapters will be provided later.

Tooling Measurements

A machinist must be able to measure how far the cutter travels. As mentioned previously, there are graduated scales on the cross-slide and compound-rest feed dials/handles, as shown in *Figure 3.12*. Although the majority of machines found in the U.S. are calibrated in inches, machines with ISO (metric) increments are slowly making their way into many shops and factories. You may even find that some machines have both measurement systems on the same dial. This is accomplished by making one scale slightly

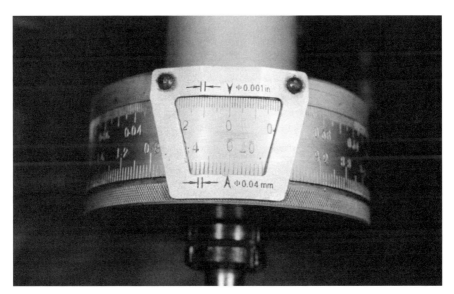

Figure 3.12 A calibrated scale on a feed wheel. On lathes with a 1:1 cutting ratio, the tool moves one-half of the indicated dial reading in order to cut the workpiece exactly to size. On lathes with a 2:1 cutting ratio, the tool moves exactly the same amount as the indicated dial reading, but removes twice as much material.

larger or smaller than the other, much the same as including speed in both miles and kilometers per hour on most modern auto speedometers.

The actual calibrations are dependent upon the machine's manufacturer. On most lathes, you will find that the cross-slide moves .100" per turn of the handle. Some machines move .250" per turn. The compound-slide is similar in movement, i.e., .100" or .200" per turn.

Feed ratios and graduated scales

Lathe feed dials have travel/cutting ratios of either 1:1 or 2:1. You will need to identify the ratio on your lathe before performing any machining operations. On 1:1 machines, the feed dial moves the cross-slide or compound one-half ($1/2$) the actual distance specified on the dial. This allows the cutter to remove material from the workpiece to the precise depth that the dial is set to.

To understand this, let's take a look at the relationship between the mounted workpiece and the cutting tool. Let's say that the workpiece is 1.000" in diameter, and let's assume that the cutting tool is just touching the workpiece. Now, start the machine and then set the graduated scale on the feed wheel for the cross-slide at .010". Allow the tool to move into the workpiece as the carriage moves forward approximately $1/4$"; then stop the carriage and turn the machine off. Use a micrometer to measure the diameter of the part where it has just been machined by the lathe—you should find that its size is .990". This machine would be considered to have a 1:1 feed ratio: for every one-thousandth that the graduated scale indicates that the cutter was fed into the workpiece, exactly one-thousandth of an inch was removed from the overall diameter of the workpiece—a ratio of 1:1.

On lathes with a 2:1 ratio, the amount of stock removed in relation to the feed indicated on the graduated scale would be twice as much. To explain, let's use the same 1.000" diameter stock. The cutter is, once again, just touching the workpiece. Turn the machine on, dial the cross-slide .010", and move the carriage forward approximately $1/4$". Then shut the machine off and measure the workpiece. The diameter will be .980", or .020" smaller than our original size (twice the amount of advance indicated on the graduated scale). This can be explained with the following example.

On 1:1 machines, when .010" is shown on the graduated scale, the cutter is removing .005" around the circumference of the workpiece, which results in the overall diameter being reduced by twice that amount, or .010". Think of it this way: if you peel an apple you are removing the skin. Once the apple has all of the skin removed, its diameter is smaller by twice the skin's thickness. Take, for example, an apple, with skin, that is 3.000" diameter. The thickness of the apple's skin is .020". The diameter of the apple contains two skin thickness, which is equal to 2 × .020" = .040". Therefore, subtract the thickness of the skin on both sides from the diameter of the apple, 3.000" – .040" = 3.960", which will be the skinless apple's diameter. Remember that when using 1:1 machines, you are reducing the workpiece diameter by exactly the dial setting. On 2:1 machines, you reduce the workpiece diameter by twice the dial setting.

Now that we understand some of the mechanics of measurement (at least with the cross-slide), let's have another look at the compound-rest. As mentioned above, the compound-rest provides a base for the toolpost assembly, and it plays a special role because of its ability to rotate and be set at different angles. When the compound-rest is set perpendicular to the ways, it is parallel to the cross-slide, resulting in a feed of 1:1 or 2:1, depending on how the machine is calibrated. Changing the angle of the compound-rest assembly to 30°, however, reduces the amount of infeed movement of the cutting tool (as indicated on compound feed handle's calibrated scale) by one-half. Furthermore, setting the compound-rest at an angle of 84° 16' will result in the removal of only 10% of the infeed movement predicted by the feed that was dialed in on the calibrated scale. Therefore, it can easily be seen that increased precision can be gained with the lathe by altering the angle of the compound-rest assembly relative to the cross-slide, as even a lathe feed scale calibrated in tenth-of-an-inch infeed increments can be used to accurately remove material in one-hundredth-of-an-inch increments.

Altering the angle of the compound-rest can also be useful in other operations, especially thread cutting, where setting the compound at 29° offers many benefits.

Indicator readouts

The readouts used to indicate linear movement may be either analog or digital. Analog measurements use dials with measurement scales inscribed

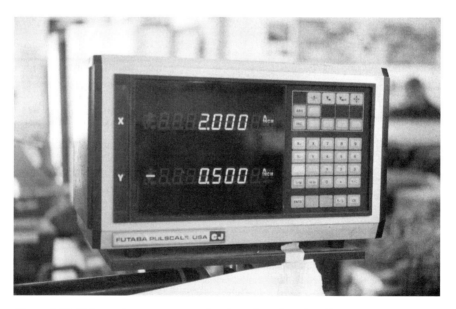

Figure 3.13 *This machine is set up with a digital readout (DRO).*

on them. Simply line up the scale with a zero or reference line/point and read the measurement. Analog type measuring scales were used on all early lathes. However, the accuracy of the lines gets a bit cloudy when the measurement falls between the lines. Is it located nearer one line or the other, or exactly in the middle of the reference lines? Sometimes it's really hard to tell.

The invention and introduction of the digital readout (DRO) has changed, forever, the accuracy of our machining measurements. If the number desired is not displayed on the dial, then the measurement is not the size you want it to be. DROs use a special encoder that is attached to the machine for reference. The reference signal is sent to the digital processor for calculation, and then sent on to the display panel. When fitted to the lathe, precise measurements can be made to within .001", .0005", or even .0002" depending upon the accuracy of the readout. Encoders can be located on the carriage to track movement to and from the spindle, and on the cross-slide to tally the distance across the face of the workpiece. Highly accurate measuring devices are not a guarantee, however, of machining accuracy. The lathe must be checked regularly to see that wear in the spindle bearings, ways, or gears has not affected the ability of the machine to perform to precision standards.

Carriage measurement indicators

When setting up the lathe for production runs, you may need to make minute adjustments to compensate for tool wear, or to make final adjustments to the carriage. This can best be accomplished with the carriage *micrometer stop*. This little device, as illustrated in *Figure 3.14*, is attached to the ways of the machine and can usually be mounted in front of or behind the carriage. It is calibrated in thousandths of an inch or hundredths of a

Figure 3.14 *The micrometer stop can be adjusted for precise stop points and to compensate for tool wear.*

Figure 3.15 *A ride-along indicator makes long measurements possible on machines not equipped with digital readouts.*

millimeter. The tool is used to very precisely limit the travel of the carriage. However, the operator must take care not to crash the carriage into the micrometer stop. This may cause the stop to move out of position and result in an inaccurate measurement.

There is a final group of what might be called *ride-along indicators* that can be attached to the carriage, and sometimes the cross-slide, to measure linear movement. The dials can be set to a zero point, which makes them very convenient when making multiple cuts from different starting points. A ride-along indicator is shown in *Figure 3.15*.

A handy tool to have in your toolbox is a dial indicator, like the one shown in *Figure 3.16*, with a magnetic back or base. The indicator can be placed on the bed of the lathe in front of (or in back of) the carriage to measure short distances (no more than the capacity of the indicator). The use of indicators with travel of 0–1" and 0–2" are especially handy for this purpose. Using a dial indicator to cut a part to exacting sizes is much more efficient than using the compound-slide feed dial.

Figure 3.16 *A dial indicator fitted with a magnetic back is a very useful tool on a lathe.*

Review Questions for Chapter 3

1. The first lathe built in America with a leadscrew was built in _____.

    A. 1740

    B. 1776

    C. 1836

    D. 1850.

2. The main difference between a *conventional* and *gap* bed lathe is _____.

    A. The size and weight

    B. A removable section of the bed

    C. The size of the face plate

    D. The gap between the machine legs.

3. Which material is best suited for the construction of a lathe?

    A. Cast iron

    B. Pig iron

    C. Steel billet

    D. Cast iron with steel wear plates.

Chapter 3 - Engine Lathes

4. Single phase (1φ) powered machines require wiring with _____.

   A. Two conductors

   B. Two conductors and one ground

   C. Two conductors and two grounds

   D. Three conductors and one ground.

5. Lathes are fitted with either a _____ or a _____ to turn them ON and OFF.

   A. Switch or a lever

   B. Switch or a knob

   C. Knob or a feed wheel

   D. Feed wheel or a switch.

6. Lathes are fitted with either _____ or _____ drive systems.

   A. Belt or pulley

   B. Belt or gear

   C. Gear or spindle

   D. Spindle or pulley.

7. There are _____ different types of spindle types used on engine lathes.

   A. 1

   B. 2

   C. 3

   D. 4.

8. Cast iron tends to hold its shape in both hot and cold weather, whereas steel has _____.

   A. A crystalline structure

   B. Poor thermal conductivity

   C. Excellent thermal conductivity

   D. None of these.

9. The quick-change gearbox is connected to the _____.

   A. Apron

   B. Cross-feed

   C. Tail stock

   D. Leadscrew.

10. The foot stock sets _____ the head stock.

    A. Beside

    B. Opposite

    C. Inside

    D. Outside.

11. The graduated scale on the tail stock is used to measure _____.

    A. The size of the spindle

    B. The maximum size of stock that can be turned on the machine

    C. How far the tail stock quill sticks out

    D. The length of the workpiece.

12. Feed dials are used to _____.

    A. Measure cross-slide movement

    B. Measure compound-slide movement

    C. Move the cross- and compound-slides

    D. All of the above.

CHAPTER 4
Toolholders

The selection and use of proper toolholders and cutting tools can be very important to the success of any cutting operation. In this chapter, we will look at several types of holders that are available for lathes, and explain how they are used in machining.

First, we must distinguish between a toolholder and a cutting tool. Toolholders are those parts or devices used to hold, carry, and/or otherwise position the cutting tool. Even though the holder has nothing to do with the actual cutting process, its use makes it possible to place the cutting tool in contact with the workpiece. Toolholders must be exceptionally rigid, yet they must be flexible enough to allow them to be positioned quickly and accurately.

Cutting tools, on the other hand, fit into toolholders, and they cut things. Cutting tools must be able to *turn*, *face*, *thread*, or *part* a workpiece. Each one of the preceding terms is similar, but they are very different in execution. Cutting tools must be made or sharpened to certain geometric angles, and these angles establish the cutting *tool geometry*. The actual *angle of attack* (the angle of contact between the tool and the workpiece) is mandated by the speed of the workpiece, depth of cut, tool geometry, and material type. We will take an in-depth look at cutting tools and tool geometry in the chapters that follow.

Toolholders come in a variety of types and sizes. Some are simple, some are complex, but all perform the same job—they hold cutting tools. The majority of cutting tools are square; sizes include: $1/8$", $3/16$", $1/4$", $5/16$", $3/8$", $1/2$", $5/8$", $3/4$", and 1". Larger sizes are available for larger machines.

Commercially manufactured toolholders are sized for machine and tool bit requirements. You would not want to use a toolholder designed for a 36" swing lathe on one that can swing only 9", or vice versa. Small lathes (6"-12" swing) will typically use $3/8$" or $1/2$" square tool bits.

Standard Toolholders

Standard toolholders used to be the tooling package shipped with every new lathe, but not any more. They mount in a slot on the lathe's tool post, and they are still a viable toolholder for certain turning and facing operations. On some tool posts, a *half-moon* rocker is used to help locate the tool at the workpiece centerline. Standardized tool posts are shown in *Figure 4.1*, and the one on your lathe should match, or be very similar to, one of the posts illustrated.

There are several standard holders available to suit the multitude of lathe operations. They are available in many configurations, including straight-shank offset, right-shank offset, and left-shank offset types. On "offset" tools, the offset is 34°. Knurling tools, boring bar, and parting

Chapter 4 - Toolholders

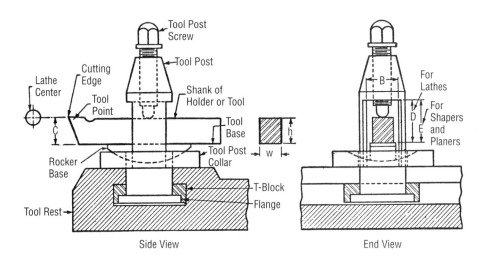

(a) Single-screw Tool-post with Rocker Base. Used on most small lathes.

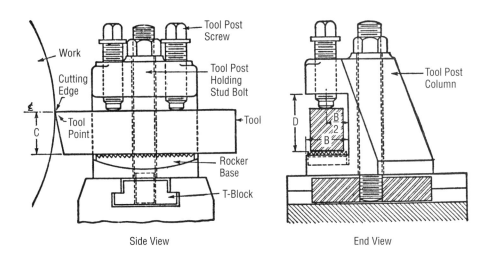

(b) Open-side Tool-post with Serrated Rocker Base.

Figure 4.1 *Standard tool posts. (ASA B5.22.)*

62   Lathe Operation and Maintenance

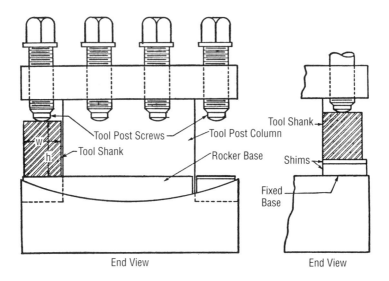

(c) Four-way Open-side Turret Tool-post.

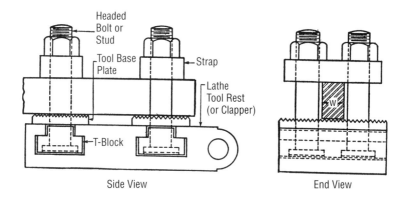

(d) Strap-and-stud Clamp Type of Toolholder with Serrated Base Used on Larger Lathes.

Figure 4.1 *(continued) Standard tool posts. (ASA B5.22.)*

toolholders round out the majority of selections, and several varieties are shown in *Figure 4.2*. As would be expected, the standard toolholders come in different sizes. *Tables 4.1 through 4.4* list dimensions of many of the most commonly used standard holders that are commercially available. The tool bits that fit these holders are covered in the next chapter.

In addition to the holders that will be discussed in this chapter, there are a wide variety of specialized holders that can be very handy for machining operations that have special requirements. You can find these listed and illustrated in catalogs published by tool supply companies.

4-position/station toolholder

Just about every lathe manufactured and shipped today has a 4-position/station toolholder (as shown in *Figure 4.3*) installed as standard equipment. The turret style (this holder is sometimes referred to as a *turret tool* post) makes it easy to operate and allows the mounting of four (4) different tools that can be selected (indexed) as required. As with other

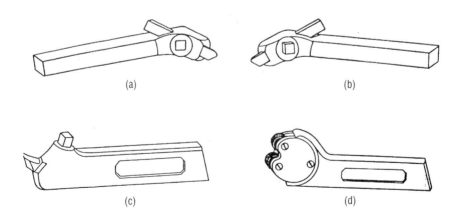

Figure 4.2 Common toolholders: (a) left-hand, (b) right-hand, (c) straight, (d) knurling tool.

Table 4.1: Standard Toolholder Dimensions.

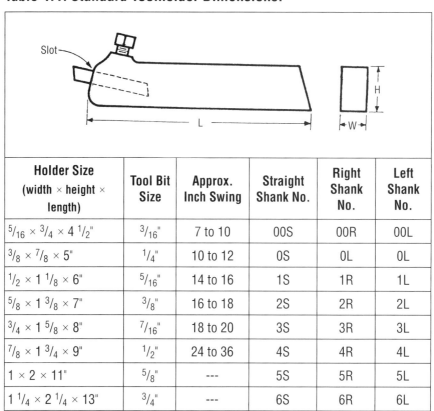

Holder Size (width × height × length)	Tool Bit Size	Approx. Inch Swing	Straight Shank No.	Right Shank No.	Left Shank No.
5/16 × 3/4 × 4 1/2"	3/16"	7 to 10	00S	00R	00L
3/8 × 7/8 × 5"	1/4"	10 to 12	0S	0L	0L
1/2 × 1 1/8 × 6"	5/16"	14 to 16	1S	1R	1L
5/8 × 1 3/8 × 7"	3/8"	16 to 18	2S	2R	2L
3/4 × 1 5/8 × 8"	7/16"	18 to 20	3S	3R	3L
7/8 × 1 3/4 × 9"	1/2"	24 to 36	4S	4R	4L
1 × 2 × 11"	5/8"	---	5S	5R	5L
1 1/4 × 2 1/4 × 13"	3/4"	---	6S	6R	6L

toolholders, one size will not fit all machines. *Table 4.5* provides dimensions of recommended size toolholders for lathes up to a 20" swing, and *Table 4.6* provides the same dimensions for heavy-duty forged steel tool post recommendations for lathes with swings as large as 48".

Table 4.2: Cut-Off Toolholder Dimensions.

Holder Size (width × height × length)	Blade Size	Approx. Inch Swing	Straight Shank No.	Right Shank No.	Left Shank No.
$5/16 \times 3/4 \times 4\ 1/2"$	$3/32 \times 1/2"$	6 to 8	19S	29R	29L
$3/8 \times 7/8 \times 5"$	$3/32 \times 5/8"$	9 to 10	20S	30R	30L
$1/2 \times 1\ 1/8 \times 6"$	$1/8 \times 3/4"$	11 to 14	21S	31R	31L
$5/8 \times 1\ 3/8 \times 7"$	$1/8 \times 7/8"$	14 to 16	22S	32R	32L
$3/4 \times 1\ 5/8 \times 8"$	$3/16 \times 1"$	16	23S	33R	33L
$7/8 \times 1\ 3/4 \times 9"$	$3/16 \times 1\ 1/8"$	18	24S	34R	34L
$1 \times 2 \times 10"$	$1/4 \times 1\ 1/4"$	20 to 24	25S	35R	35L
$1\ 1/4 \times 2\ 1/4 \times 11"$	$1/4 \times 1\ 3/8"$	30 to 36	26S	---	---

Table 4.3: Carbide Toolholder Dimensions.

Holder Size (width × height × length)	Tool Bit Size	Approx. Inch Swing	Straight Shank No.	Right Shank No.	Left Shank No.
$5/16 \times 3/4 \times 4\ 1/2"$	$1/4"$	9 to 10	T0S	T0R	T0L
$3/8 \times 7/8 \times 5"$	$5/16"$	11 to 14	T1S	T1R	T1L
$1/2 \times 1\ 1/8 \times 6"$	$3/8"$	14 to 16	T2S	T2R	T2L
$5/8 \times 1\ 3/8 \times 7"$	$7/16"$	16	T3S	TR	T3L
$3/4 \times 1\ 5/8 \times 8"$	$1/2"$	18	T4S	T4R	T4L
$7/8 \times 1\ 3/4 \times 9"$	$5/8"$	20 to 24	T5S	T5R	T5L

Table 4.4: Boring Bar Toolholder Dimensions.

Holder Size (width × height × length)	Boring Bar Diameter	Tool Bit Size	Approx. Inch Swing	Hole-to-Center Height
$5/16 \times 3/4 \times 4\ 7/8"$	$1/2"$	$3/16"$	6 to 8	$3/4"$
$3/8 \times 7/8 \times 5\ 3/8"$	$5/8"$	$3/16"$	9 to 10	$7/8"$
$3/8 \times 1\ 1/8 \times 4\ 7/8"$	$3/4"$	$1/4"$	9 to 10	$7/8"$
$1/2 \times 1\ 1/8 \times 6\ 1/8"$	$3/4"$	$1/4"$	11 to 14	$1\text{-}1/8"$
$7/16 \times 1\ 5/16 \times 5\ 7/8"$	$3/4"$	$5/16"$	11 to 14	$1\text{-}1/8"$
$5/8 \times 1\ 3/8 \times 7\ 7/16"$	$1"$	$5/16"$	14 to 16	$1\text{-}1/4"$
$3/4 \times 1\ 5/8 \times 8\ 7/16"$	$1\text{-}1/4"$	$3/8"$	16	$1\text{-}1/2"$

Table 4.5: Standard Turret Toolholders.

Dimension Range	Lathe Swing	Tool Size Range	Size (Square)	Thickness
$1\ 3/8" - 1\ 1/2"$	6" to 12"	$1/4 - 3/8"$	$2\ 1/2"$	$1\ 3/8"$
$1\ 5/8" - 1\ 13/16"$	10" to 13"	$3/8 - 1/2"$	$3\ 1/2"$	$1\ 3/4"$
$1\ 15/16" - 2\ 5/16"$	13" to 16"	$1/2 - 3/4"$	$4\ 1/2"$	$2\ 1/4"$
$2\ 1/2" - 2\ 3/4"$	14" to 20"	$7/8 - 1\ 1/4"$	$6"$	$2\ 3/4"$

Table 4.6: Forged Turret Toolholders.

Dimension Range	Lathe Swing	Tool Size Range	Size (Square)	Thickness
$1\ 5/8" - 2\ 1/2"$	13" to 20"	$3/4" - 1"$ #1	$4\ 1/2"$	$3"$
$2" - 3\ 3/4"$	16" to 24"	$1" - 1\ 1/2"$ #2 or #3	$6\ 1/2"$	$3\ 3/4"$
$2\ 3/8" - 3\ 13/16"$	18" to 32"	$1\ 1/4" - 1\ 3/4"$ #3 or #4	$7"$	$4\ 1/4"$
$2\ 7/8" - 4"$	20" to 48"	$1\ 1/2"$ #2 #3 #4 or #5	$6"$	$2\ 3/4"$

Chapter 4 - Toolholders   **67**

Figure 4.3 Installation of a 4-position toolholder is a good first step to increase part production.

Quick-change toolholders

One of the most important innovations ever introduced for the lathe was quick-change tooling. There are several variations or interpretations as to how the tools are held and changed, but all perform the same basic functions.

The system is comprised of a body, or master, holder tool post, and several tool blocks or bars. The tool blocks will typically have a dovetail cut into them to serve as the mount, and a variety of holes and/or slots that allow for secure holding of the cutting tools. As would be expected, there are different sizes of bodies and masters available for different size lathes. Some bodies are made in the turret style and can index four different toolholders for four different machining operations.

Although the tool blocks will have holes and slots cut into them to hold the tool bits, their location and positions are varied. Tool blocks include turning and facing combination bars, parting and turning, extension bars, knurling bars, master boring bar holders, and carbide insert bars.

Specialty bars that allow indicators to be attached to round stalks, and 5-C collet holders, are also available. We will discuss collets in more detail in Chapter 6.

Once you use quick-change tooling, it's pretty hard to go back to standard toolholders. After initial setup has been performed, the holders always return to their original position, tool change after tool change. Quick-change toolholders speed up the production of parts—a benefit that makes them especially desirable.

Tail Stock

The tail stock, or foot stock, has many uses when machining parts. It is used to drill stock, provide support for a long workpiece (as shown in *Figure 4.4*), and allow the part to be *offset* for taper turning.

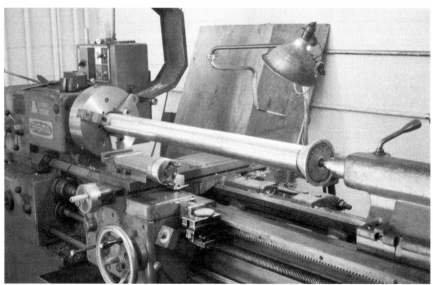

Figure 4.4 *The tail stock provides support when turning long parts.*

As described in the previous chapter, the tail stock is fitted with a Morse taper holder. The fact that the taper is less than 10° allows the tools to lock into place, but also permits easy removal. Simply crank the tail stock feed wheel counterclockwise (CCW) with some gusto, and the tool will pop out into your hand. Standard Morse taper dimensions are given in *Table 4.7*.

Morse taper drills, for example, are available in a wide variety of taper sizes and drill diameters. To allow the use of smaller diameter drills with smaller tapers, you will find adapters (sleeves) available for the task. If your lathe has a #3 MT and you want to install a $^1/_2$" diameter drill with a #2 MT taper, use a #2 to #3 Morse taper sleeve. Or, if you want to use a #4 MT drill in a #3 MT, use an extension socket for this purpose. Consult your tool supplier for a full listing of commercially available adapters and extension sockets that will fit your lathe.

Tail stock mounted indexer

To increase the speed of the standard lathe, you might want to consider the installation and use of a tail stock indexer. This tool, as shown in *Figure 4.5*, allows the mounting of up to six tools in the tail stock. Though not quite as sturdy as a full turret lathe, this indexer will allow you to increase your production. The majority of those indexers made have 1" diameter holes bored into them. You must make or purchase 1" OD bushings to suit the tool diameter. Tools are held in place with a set-screw. Be sure that the tail stock is centered properly before you install the indexer.

Drill Chucks

The drill chuck is one of *the* essential tools in any machine shop, especially on a lathe. They usually have three jaws to allow for quick installation

Table 4.7: Morse Taper Dimensions.

Dimension		Drawing Symbol	Morse Taper Number								
			0	1	2	3	4	4 1/2	5	6	7
shank	Plug small end dia.	D	.25200	.36900	.57200	.77800	1.02000	1.26600	1.47500	2.11600	2.75000
	Dia. at end of socket	A	.35610	.47500	.70000	.93800	1.23100	1.50000	1.74800	2.49400	3.27000
	Length	B	2 11/32	2 9/16	3 1/8	3 7/8	4 7/8	5 3/8	6 1/8	8 9/16	11 5/8
	Depth	S	2 7/32	2 7/16	2 15/16	3 11/16	4 5/8	5 1/8	5 7/8	8 1/4	11 1/4
	Depth of drilled hole	G	2 1/16	2 3/16	2 21/32	3 5/16	4 3/16	4 5/8	5 5/16	7 13/32	10 5/32

(Continued)

Table 4.7: Morse Taper Dimensions. *(Continued)*

	Dimension	Drawing Symbol	Morse Taper Number									
			0	1	2	3	4	4 1/2	5	6	7
Shank	Depth of reamed hole	H	2 1/32	2 5/32	2 39/64	3 1/4	4 1/8	4 9/16	5 1/4	7 21/64	7 1/4
Shank	Standard plug depth	P	2	2 1/8	2 9/16	3 3/16	4 1/16	4 1/2	5 3/16	7 1/4	10
Tang	Thickness	t	5/32	13/64	1/4	5/16	15/32	9/16	5/8	3/4	1 1/8
Tang	Length	T	1/4	3/8	7/16	9/16	5/8	11/16	3/4	1 1/8	1 3/8
Tang	Radius	R	5/32	3/16	1/4	9/32	5/16	3/8	3/8	1/2	3/4
Tang	Radius	a	3/64	3/64	1/16	5/64	3/32	1/8	1/8	5/32	3/16
Slot	Width	W	11/64	7/32	17/64	21/64	31/64	37/64	21/32	25/32	1 5/32
Slot	Length	L	9/16	3/4	7/8	1 3/16	1 1/4	1 3/8	1 1/2	1 3/4	2 5/8
Slot	End of socket to slot	K	1 15/16	2 1/16	2 1/2	3 1/16	3 7/8	4 5/16	4 15/16	7	9 1/2
	Taper per inch		.052050	.049882	.049951	.050196	.051938	.052000	.052626	.052138	.052000
	Taper per foot		.62460	.59858	.59941	.60235	.62326	.62400	.63151	.62565	.62400

Figure 4.5 A tail stock mounted indexer.

Figure 4.6 A "keyless" chuck. The jaws are tightened and loosened by hand tightening.

and removal of drill bits for drilling operations, and provide a relatively high degree of support necessary for precision work.

Drill chucks are tightened with a key, or by hand (keyless chuck, as shown in *Figure 4.6*), to clamp the cutting tool. Chucks are normally mounted onto a *straight-shank* and held with a toolholder on the compound-rest, or with a Morse taper adapter. Sizes are determined by the maximum amount that the chuck jaws will open. The majority of drill chucks used on lathes will hold at least a $1/2$" diameter tool. A $5/8$" diameter capacity chuck would be a better option because a tool just slightly larger than the $1/2$" limit is often needed for drilling.

When purchasing a drill chuck, two questions will need to be answered before the acquisition: 1) what will the capacity be? and 2) what size adapter will be needed to mount the chuck? Chucks are manufactured in many sizes ranging from completely closed (0), to a maximum opening of approximately

Figure 4.7 *A drill chuck in operation.*

1" (25.4 mm). The majority of drill chucks are mounted onto a tapered shaft that conforms to the Jacob's standard taper (JT) and are found in the following sizes: 0-JT, 1-JT, 2-JT, 3-JT, 4-JT, 5-JT, 6-JT, and the 33-JT (you may sometimes hear drill chucks referred to as "Jacob's chucks," an early brand name for these workholding devices). A straight or Morse taper is found on the machine mounting end.

The chuck closure mechanism will use a plain or ball-thrust bearing. The ball bearing type is designed for production drilling operations and reduces friction in the jaw closing mechanism. As a result, the operator can apply greater gripping force onto the drill or tool bit shank.

Chuck keys used to tighten the drill will be mated to the particular size chuck you use. Do not attempt to fit a key from one size chuck to another. You will only destroy the key and/or the teeth on the chuck. Also, be sure to remove the key before starting the lathe. When finished using the drill chuck, insert the key into the jaws and tighten it hand-tight; this is so you don't have to waste time looking for it. Some machinists attach a cord to the key, which can then be tied or bolted to the machine. This is not a bad idea, unless you use several different sizes of chucks on the same machine.

Keyless chucks are nice because you don't have to worry about losing the chuck key. By design, their gripping power increases in proportion to the torque (power) requirement of the drilling operation.

Indexable insert toolholder

One type of toolholder we have not covered in this chapter is the holder for indexable carbide tools. They come in a wide range of styles, each designed to accommodate the shape of a specific insert, and intended for a specific cutting operation. This type of tooling is used almost universally on

CNC lathes and machining centers, but less frequently on traditional engine lathes. However, you should be familiar with the seven basic styles that are shown in *Figure 4.8*. Briefly, these seven styles are as follows (of these styles, A, B, F, and G are most often used).

A style: Straight shank with 0° side cutting edge angle, used for turning.

B style: Straight shank with 15° side cutting edge angle, used for turning.

C style: Straight shank with 0° end cutting edge angle, for cut-off and grooving operations.

D style: Straight shank with 45° side cutting edge angle, for turning operations.

E style: Straight shank with 30° side cutting edge angle, for threading operations.

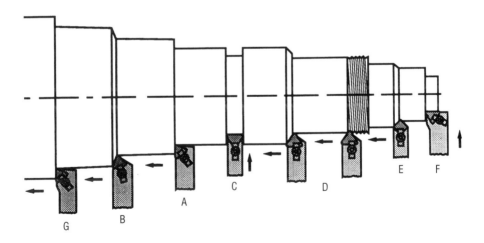

Figure 4.8 The seven basic styles (but by no means every available style) of indexable toolholders. (Courtesy, Kennametal Inc.)

F style: Offset shank with 0° end cutting edge angle, for facing operations.

G style: Offset shank with 0° side cutting edge angle. Very similar to the A style, but offset provides additional clearance for machining close to the lathe chuck.

Indexable toolholders are available as right-hand or left-hand tools. The problem of identifying a right-hand or left-hand tool can be resolved if you remember the following. When holding the shank of a right-hand tool with the insert facing upward (as shown in *Figure 4.9*), it will cut from right to left. The left-hand tool, when held as shown in the same figure, will cut from left to right.

Another toolholder that is available in both right-hand and left-hand styles is the holder used for brazed single point carbide tools. They have not been discussed in this chapter because the tool bit and cutter are integrated

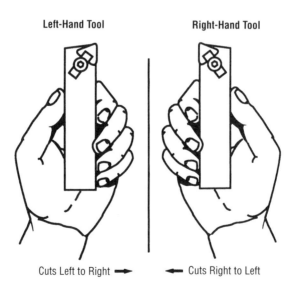

Figure 4.9 Left-hand and right-hand tooling. (Courtesy, Kennametal Inc.)

into a single unit. Since the cutting tool incorporates the holder, they will be discussed in the next chapter on cutting tools.

Center Drill Drivers

Center drill drivers are used to hold center drills in a Morse taper. They may be used in either the head or tail stock. The drivers are made of steel for durability. The outside of the driver is a Morse taper shape in size #1, #2, or #3, and the hole in the middle is drilled to fit center drill sizes from #1 to #8. It is also possible to hold other cutting tools with round shanks in the driver but some experimenting may be required.

Table 4.8: Center Drill Drivers.

Center Drill Size	Body Diameter	Morse Taper #1	Morse Taper #2	Morse Taper #3
1, 11	1/8"	✔		
2, 12	3/16"	✔	✔	
3, 13	1/4"	✔	✔	✔
4, 14	5/16"	✔	✔	
5, 15	7/16"		✔	
6, 16	1/2"		✔	
7, 17	5/8"			✔
8, 18	3/4"			✔

Review Questions for Chapter 4

1. The majority of cutting tools are _____.

    A. Round

    B. Hexagon

    C. Octagon

    D. Square.

2. A right-handed toolholder cuts from _____.

    A. Right to left

    B. Left to right

    C. Straight

    D. None of these.

3. If the tail stock is *offset*, the machine will cut _____ _____.

    A. A straight and round part

    B. A groove in the workpiece

    C. A taper

    D. An internal hole.

4. A tail stock mounted indexer will increase _____ _____.

   A. The number of tail stocks

   B. Production

   C. The number of speeds and feeds

   D. The number of threads that the machine can cut.

5. Drill chucks are typically held in place with a _____.

   A. Morse bevel

   B. Morse taper

   C. Collar and set screw

   D. None of these.

6. Indexable insert toolholders are the latest generation of toolholders used in production machine shops.

   A. True

   B. False.

Chapter 5: Cutting Tools

Cutting tools must be made from materials harder than the workpiece that is to be machined. Relatively soft materials, such as aluminum, can be cut with standard (and therefore less costly) tooling. Cutting exotic and space age materials often requires the use of cutting tools and holders more costly than the job itself.

High-speed steel tools

The majority of cutting tools are made from special steel alloys or hybrid compositions. Among the most common materials used for lathe tools are *high-speed tool steel* (HSS) and *carbide*. High-speed tool steels contain large amounts of molybdenum (Mo), chromium (Cr), and/or vanadium (V), combined with a basic fortifier—carbon (C). These steels can be divided into two groupings: tungsten (W) and molybdenum. Tungsten tool steels will have a prefix *T* before the number, which designates its grade.

Molybdenum alloyed tool steels will have an *M* prefix.

Tungsten HSS has been replaced almost entirely by the molybdenum (moly) based materials for cutting tools. The most common of these moly steels in use today are M2 and M40. The hardness of these steels will range between 63 R_C and 70 R_C. (Hardness of tool steels is usually measured with the Rockwell "C" scale, indicated by the symbol "R_C.") In general, high-speed steels will retain their effective hardness at temperatures ranging up to 1000° and 1100° F, which is high enough to maintain sufficient hardness during most lathe operations.

Most tools made from HSS are ground from solid blanks or bars and contain a single cutting edge (hence the name "single-point tools"). Dimensions of standard high-speed steel single point tools are given in *Table 5.1*, and for the majority of lathe operations these tools will provide very satisfactory results. In addition, they can be resharpened numerous times if used properly, and therefore are capable of providing a very long service life.

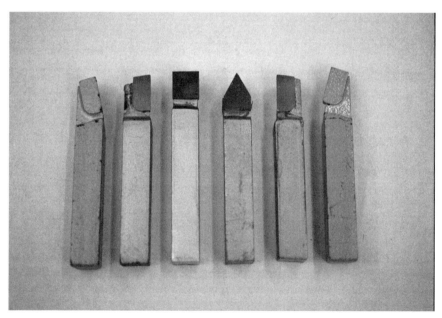

Figure 5.1 *A selection of cutting tools used on a lathe.*

Table 5.1: Dimensions of American National Standard High-Speed Steel Single-Point Tools.

Squares		Flats		
Size (inches)		Size (inches)		
Cross-section A = B	Length C	Width A	Height B	Length C
3/16	2 1/2	1/4	3/8	3
1/4	2 1/2	1/4	1/2	4
5/16	2 1/2	1/4	3/4	5
3/8	3	1/4	1	7
7/16	3 1/2	5/16	7/16	3 1/2
1/2	4	5/16	5/8	4 1/2
9/16	4	3/8	1/2	4
5/8	4 1/2	3/8	5/8	4 1/2
3/4	5	3/8	3/4	5
7/8	6	3/8	1	6
1	7	1/2	5/8	4 1/2
1 1/8	7	1/2	3/4	5
1 1/4	9	1/2	1	7
1 1/2	12	1/2	1 1/4	6
–	–	5/8	3/4	5
–	–	5/8	7/8	6
–	–	5/8	1, 1 1/4, or	7
–	–	3/4	1, 1 1/4, or	7
–	–	7/8	1 1/4	7
–	–	1	1 1/2	7

Source, ANSI B94.10-1967, as published by the American Society of Mechanical Engineers.

Brazed single-point cutting tools

Single-point tools may be made of high-speed steel, carbon steel, cobalt alloy, or carbide, and they are among the most popular cutting tools used on the engine lathe. However, due to the expense of producing a complete tool from costly materials, tools made from one piece of solid material have largely been replaced with brazed-tipped tools. Typically, brazed-tip tools are comprised of a shank of relatively inexpensive material with a tip or blank of more costly cutting material brazed onto the cutting end. Materials used for the tip include high-speed steel, carbide, and cubic boron nitride.

Carbide is a sintered (a high-temperature process that binds the material) metal product made from highly refined tungsten ores that are approximately 94% to 97% pure. They retain their hardness at temperatures up to around 1400° F. Carbide has a hardness comparable to 90.5 R_C to 93.5 R_C, and can be ground to a fine edge, making it a very desirable cutting tool material. The big plus for carbide tooling is that it can cut many materials at speeds up to four times faster than HSS. It is also a prerequisite for cutting certain space-age alloys. Carbide tools are less tolerant to shock loading than HSS, which means that they cannot be dropped or banged around without damaging the cutting edge. Carbide is easy to distinguish from HSS: it does not have a bright, shiny appearance (it is dull gray in color), and it is quite heavy by comparison.

Brazed single-point turning tools are available in standard sizes designated by the letters A to G. Offset point styles A, B, E, F, and G are available as either right- or left-hand cutting tools. For square shanks, the number following the letter designation indicates the number of sixteenths of an inch required to equal the height or width of the tool. For rectangular shanked tools, the first number is the total of eighths of an inch in the shank width, and the second number is the total of fourths of an inch in the shank height. The nose radius is related to shank size, and care should be taken to match the tool size to the finish requirements of each machining operation. Dimensions of standard brazed single-point carbide tipped tools are given in *Tables 5.2 – 5.8*, and standard replacement tip dimensions are given in *Table 5.9*.

Table 5.2: Style A (0° Side Cutting-edge Angle) Brazed Single-point Tools. *For Turning, Facing, or Boring to 90° Square Shoulder. (Courtesy Kennametal Inc.)*

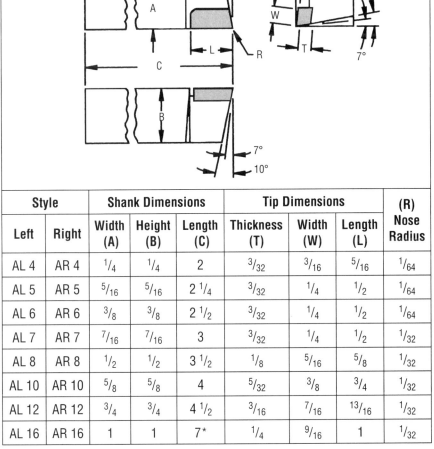

Style		Shank Dimensions			Tip Dimensions			(R) Nose Radius
Left	Right	Width (A)	Height (B)	Length (C)	Thickness (T)	Width (W)	Length (L)	
AL 4	AR 4	$1/4$	$1/4$	2	$3/32$	$3/16$	$5/16$	$1/64$
AL 5	AR 5	$5/16$	$5/16$	$2\,1/4$	$3/32$	$1/4$	$1/2$	$1/64$
AL 6	AR 6	$3/8$	$3/8$	$2\,1/2$	$3/32$	$1/4$	$1/2$	$1/64$
AL 7	AR 7	$7/16$	$7/16$	3	$3/32$	$1/4$	$1/2$	$1/32$
AL 8	AR 8	$1/2$	$1/2$	$3\,1/2$	$1/8$	$5/16$	$5/8$	$1/32$
AL 10	AR 10	$5/8$	$5/8$	4	$5/32$	$3/8$	$3/4$	$1/32$
AL 12	AR 12	$3/4$	$3/4$	$4\,1/2$	$3/16$	$7/16$	$13/16$	$1/32$
AL 16	AR 16	1	1	7*	$1/4$	$9/16$	1	$1/32$

* AL 16 and AR 16 also available in 6" length.

Table 5.3: Style B (15° Side Cutting-edge Angle) Brazed Single-point Tools. *For Turning, Facing, or Boring with Lead Angle in Order to Allow Higher Feeds with Reduced Pressure. (Courtesy Kennametal Inc.)*

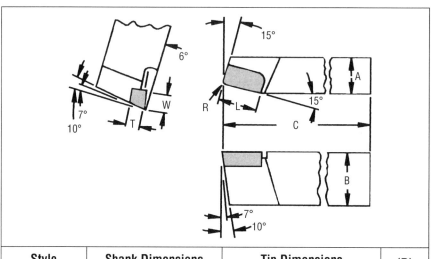

Style		Shank Dimensions			Tip Dimensions			(R) Nose Radius
Left	Right	Width (A)	Height (B)	Length (C)	Thickness (T)	Width (W)	Length (L)	
BL 4	BR 4	$1/4$	$1/4$	2	$1/16$	$5/32$	$1/4$	$1/64$
BL 5	BR 5	$5/16$	$5/16$	$2\,1/4$	$3/32$	$3/16$	$5/16$	$1/64$
BL 6	BR 6	$3/8$	$3/8$	$2\,1/2$	$3/32$	$1/4$	$1/2$	$1/64$
BL 7	BR 7	$7/16$	$7/16$	3	$3/32$	$1/4$	$1/2$	$1/32$
BL 8	BR 8	$1/2$	$1/2$	$3\,1/2$	$1/8$	$5/16$	$5/8$	$1/32$
BL 10	BR 10	$5/8$	$5/8$	4	$5/32$	$3/8$	$3/4$	$1/32$
BL 12	BR 12	$3/4$	$3/4$	$4\,1/2$	$3/16$	$7/16$	$13/16$	$1/32$
BL 16	BR 16	1	1	7*	$1/4$	$9/16$	1	$1/32$

* AL 16 and AR 16 also available in 6" length.

Table 5.4: Style C (0° End Cutting-edge Angle) Brazed Single-point Tools. *For Chamfering, Facing, and Plunge Cutting. (Courtesy Kennametal Inc.)*

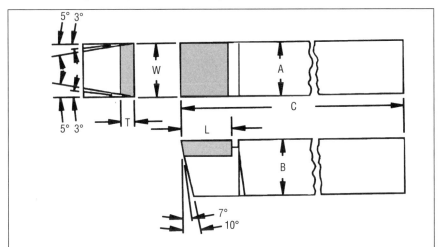

Style	Shank Dimensions			Tip Dimensions		
	Width (A)	Height (B)	Length (C)	Thickness (T)	Width (W)	Length (L)
C 4	1/4	1/4	2	1/16	1/4	5/16
C 5	5/16	5/16	2 1/4	3/32	5/16	3/8
C 6	3/8	3/8	2 1/2	3/32	3/8	3/8
C 7	7/16	7/16	3	3/32	7/16	1/2
C 8	1/2	1/2	3 1/2	1/8	1/2	1/2
C 10	5/8	5/8	4	5/32	5/8	5/8
C 12	3/4	3/4	4 1/2	3/16	3/4	3/4
C 16	1	1	7*	1/4	1	3/4
C 20	1 1/4	1 1/4	8**	5/16	1 1/4	3/4
C 44	1/2	1	7*	3/16	1/2	1/2
C 54	5/8	1	6	1/4	5/8	5/8
C 55	5/8	1 1/4	8**	1/4	5/8	5/8

* Also available in 6" length. **Also available in 7" length.

Table 5.5: Style D (Cut-off) Brazed Single-point Tools.

For Right-hand or Left-hand Turning, Boring, or Chamfering.
(Courtesy Kennametal Inc.)

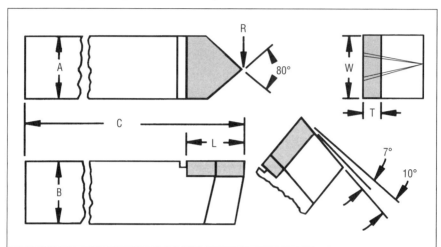

Style	Shank Dimensions			Tip Dimensions			(R) Nose Radius
	Width (A)	Height (B)	Length (C)	Thickness (T)	Width (W)	Length (L)	
D 4	1/4	1/4	2	1/16	1/4	5/16	1/64
D 5	5/16	5/16	2 1/4	3/32	5/16	3/8	1/64
D 6	3/8	3/8	2 1/2	3/32	3/8	1/2	1/64
D 7	7/16	7/16	3	3/32	7/16	1/2	1/32
D 8	1/2	1/2	3 1/2	1/8	1/2	1/2	1/32
D 10	5/8	5/8	4	5/32	5/8	5/8	1/32
D 12	3/4	3/4	4 1/2	3/16	3/4	3/4	1/32
D 16	1	1	7*	1/4	1	3/4	1/32

* D 16 also available in 6" length.

Table 5.6: Style E (60° Nose-angle) Brazed Single-point Tools.

For 60° V-threading. (Courtesy Kennametal Inc.)

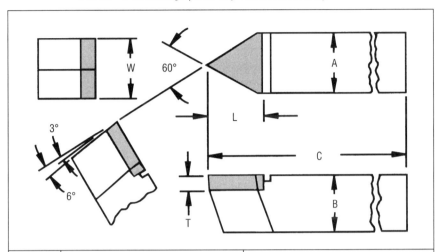

Style	Shank Dimensions			Tip Dimensions		
	Width (A)	Height (B)	Length (C)	Thickness (T)	Width (W)	Length (L)
E 5	$5/16$	$5/16$	$2\,1/4$	$3/32$	$5/16$	$3/8$
E 6	$3/8$	$3/8$	$2\,1/2$	$3/32$	$3/8$	$1/2$
E 8	$1/2$	$1/2$	$3\,1/2$	$1/8$	$1/2$	$1/2$
E 10	$5/8$	$5/8$	4	$5/32$	$5/8$	$5/8$
E 12	$3/4$	$3/4$	$4\,1/2$	$3/16$	$3/4$	$3/4$

Table 5.7: Style F (0° End Cutting-edge Angle Offset) Brazed Single-point tools. *For Facing (Tool Shank 90° to the Work Axis) or Turning and Boring (Tool Shank Parallel to the Work Axis). (Courtesy Kennametal Inc.)*

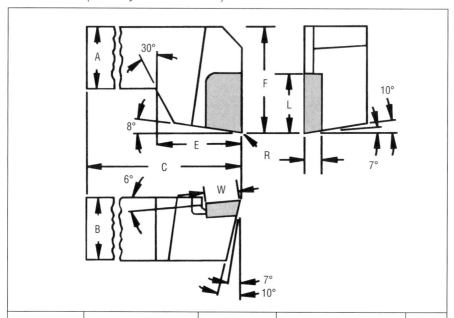

Style		Shank Dimensions			Head Dimensions		Tip Dimensions			(R) Nose Radius
Left	Right	Width (A)	Height (B)	Length (C)	Length (E)	Height (F)	Thickness (T)	Width (W)	Length (L)	
FL 8	FR 8	$1/2$	$1/2$	$3\ 1/2$	$3/4$	$3/4$	$1/8$	$5/16$	$5/8$	$1/32$
FL 10	FR 10	$5/8$	$5/8$	4	1	1	$5/32$	$3/8$	$3/4$	$1/32$
FL 12	FR 12	$3/4$	$3/4$	$4\ 1/2$	$1\ 1/8$	$1\ 3/8$	$3/16$	$7/16$	$13/16$	$1/32$
FL 16	FR 16	1	1	7*	$1\ 3/8$	$1\ 3/4$	$1/4$	$9/16$	1	$1/32$

* Also available in 6" length.

Table 5.8: Style G (0° Side Cutting-edge Angle Offset) Brazed Single-point tools. *For Turning a Shoulder (Tool Shank 90° to the Work Axis). (Courtesy Kennametal Inc.)*

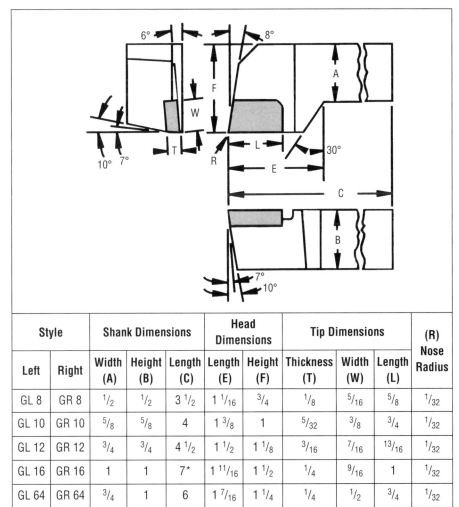

Style		Shank Dimensions			Head Dimensions		Tip Dimensions			(R) Nose Radius
Left	Right	Width (A)	Height (B)	Length (C)	Length (E)	Height (F)	Thickness (T)	Width (W)	Length (L)	
GL 8	GR 8	1/2	1/2	3 1/2	1 1/16	3/4	1/8	5/16	5/8	1/32
GL 10	GR 10	5/8	5/8	4	1 3/8	1	5/32	3/8	3/4	1/32
GL 12	GR 12	3/4	3/4	4 1/2	1 1/2	1 1/8	3/16	7/16	13/16	1/32
GL 16	GR 16	1	1	7*	1 11/16	1 1/2	1/4	9/16	1	1/32
GL 64	GR 64	3/4	1	6	1 7/16	1 1/4	1/4	1/2	3/4	1/32

* Also available in 6" length.

Table 5.9: Sizes and Catalog Numbers for Replacement Tips for Brazed Single-point Tools.

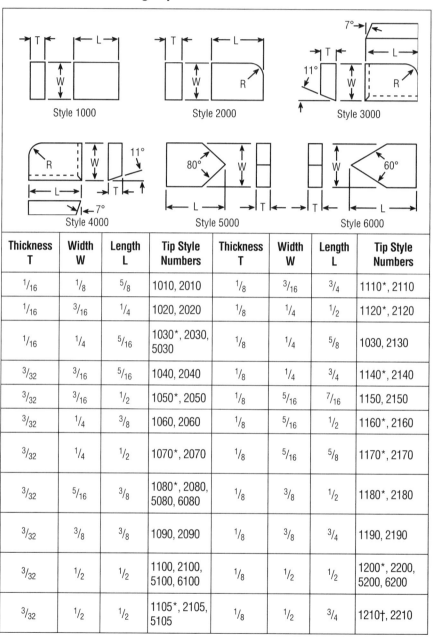

Thickness T	Width W	Length L	Tip Style Numbers	Thickness T	Width W	Length L	Tip Style Numbers
$1/16$	$1/8$	$5/8$	1010, 2010	$1/8$	$3/16$	$3/4$	1110*, 2110
$1/16$	$3/16$	$1/4$	1020, 2020	$1/8$	$1/4$	$1/2$	1120*, 2120
$1/16$	$1/4$	$5/16$	1030*, 2030, 5030	$1/8$	$1/4$	$5/8$	1030, 2130
$3/32$	$3/16$	$5/16$	1040, 2040	$1/8$	$1/4$	$3/4$	1140*, 2140
$3/32$	$3/16$	$1/2$	1050*, 2050	$1/8$	$5/16$	$7/16$	1150, 2150
$3/32$	$1/4$	$3/8$	1060, 2060	$1/8$	$5/16$	$1/2$	1160*, 2160
$3/32$	$1/4$	$1/2$	1070*, 2070	$1/8$	$5/16$	$5/8$	1170*, 2170
$3/32$	$5/16$	$3/8$	1080*, 2080, 5080, 6080	$1/8$	$3/8$	$1/2$	1180*, 2180
$3/32$	$3/8$	$3/8$	1090, 2090	$1/8$	$3/8$	$3/4$	1190, 2190
$3/32$	$1/2$	$1/2$	1100, 2100, 5100, 6100	$1/8$	$1/2$	$1/2$	1200*, 2200, 5200, 6200
$3/32$	$1/2$	$1/2$	1105*, 2105, 5105	$1/8$	$1/2$	$3/4$	1210†, 2210

(Continued)

Table 5.9: Sizes and Catalog Numbers for Replacement Tips for Brazed Single-point Tools. *(Continued)*

Thickness T	Width W	Length L	Tip Style Numbers	Thickness T	Width W	Length L	Tip Style Numbers
$1/8$	$3/4$	$3/4$	1215, 2215	$1/4$	$3/8$	$3/4$	1360, 3360, 4360
$5/32$	$3/8$	$9/16$	1220, 2220	$1/4$	$7/16$	$5/8$	1370, 3370, 4370
$5/32$	$3/8$	$3/4$	1230, 2230	$1/4$	$1/2$	$3/4$	1380*, 3380, 4380
$5/32$	$5/8$	$5/8$	1240*, 2240, 5240, 6240	$1/4$	$9/16$	1	1390, 3390, 4390
$3/16$	$5/16$	$7/16$	1250, 2250	$1/4$	$5/8$	$5/8$	1400*, 3400, 4400
$3/16$	$5/16$	$5/8$	1260*, 2260	$1/4$	$3/4$	$3/4$	1405*, 3405, 4405
$3/16$	$3/8$	$1/2$	1270*, 2270	$1/4$	$3/4$	1	1410, 3410, 4410, 5410
$3/16$	$3/8$	$5/8$	1280, 2280	$5/16$	$7/16$	$5/8$	1420, 3420, 4420
$3/16$	$3/8$	$3/4$	1290, 2290	$5/16$	$7/16$	$15/16$	1430, 3430, 4430
$3/16$	$7/16$	$5/8$	1300, 2300	$5/16$	$1/2$	$3/4$	1440*, 3440, 4440
$3/16$	$7/16$	$13/16$	1310, 2310	$5/16$	$1/2$	1	1450, 3450, 4450
$3/16$	$1/2$	$1/2$	1320*, 2320	$5/16$	$5/8$	1	1460*, 3460, 4460
$3/16$	$1/2$	$3/4$	1330†, 2330	$5/16$	$3/4$	$3/4$	1470*, 3470, 4470
$3/16$	$3/4$	$3/4$	1340*, 2340, 5340, 6340	$5/16$	$3/4$	1	1475†, 3475, 4475
$1/4$	$3/8$	$9/16$	1350, 3350, 4350	$5/16$	$3/4$	$1\ 1/4$	1480†, 3480, 4480

(Continued)

Table 5.9: Sizes and Catalog Numbers for Replacement Tips for Brazed Single-point Tools. *(Continued)*

Thickness T	Width W	Length L	Tip Style Numbers	Thickness T	Width W	Length L	Tip Style Numbers
3/8	1/2	1	1500, 3500, 4500	3/8	3/4	1 1/2	1525, 3525, 4525
3/8	5/8	1	1510†, 3500, 4500	1/2	3/4	1 1/4	1540†, 3540, 4540
3/8	3/4	1 1/4	1520†, 3520, 4520	1/2	3/4	1 1/2	1550, 3550, 4550

* These replacement tips are 1/32" oversize on width. † These replacement tips are 1/32" on length. Excess material is provided on these tips to allow the grinding of opposite tip surfaces to size. Source, ASA B5.22.

Indexable insert tooling

Production shops will almost always use tooling made with either carbide tips, or they will use replaceable cutting *inserts*. Inserts are rigidly secured to a toolholder with a screw or clamp, and they can be easily removed and rotated (indexed) on the toolholder to provide a new cutting surface in the exact same location. Because inserts are small, their replacement cost can be considerably less than that of a carbide-tipped tool, especially as each insert provides several cutting edges.

Another major advantage of inserts is that they are available in a vast number of shapes (see *Figures 5.2 and 5.3*), some of which are in the exact profile of screw threads. This makes threading operations easier and more reliable.

Inserts also come in a wide variety of grades. For example, one manufacturer (Kennametal) makes inserts for lathes in four basic groups of materials: tungsten carbide, cermet, ceramic, and polycrystalline. Each group contains a variety of cutting grades, including the following.

Chapter 5 - Cutting Tools   **95**

Figure 5.2 These carbide inserts are typical of those used in a production machine shop.

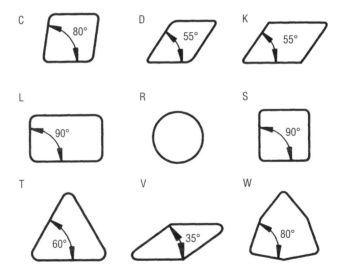

Figure 5.3 Basic insert shapes commonly used for lathe operations. C = rhombic (diamond), 80° nose angle; D = rhombic (diamond), 55° nose angle; K = parallelogram, 55° nose angle; L = rectangular, 90° nose angle; R = round; S = square, 90° nose angle; T = triangle, 60° nose angle; V = rhombic (diamond), 35° nose angle; W = trigon (hexagon), 80° nose angle.

Tungsten carbide. These are available in both uncoated and coated grades. Coatings may contain many materials and are applied with processes known as chemical vapor deposition (CVD), or physical vapor deposition (PVD). Kennametal offers four grades of uncoated carbides, nine grades of CVD coated carbides, and three grades of PVD coated carbides.

Cermet. Cermets are comprised mostly of titanium carbide (TiC) and titanium nitride (TiN) held together with a metal binder. The name comes from the CERamic METallic binders used in their manufacture.

Ceramic. Ceramic inserts can be grouped into two basic families: alumina based (aluminum oxide), and silicon-nitride based (sialon) ceramics.

Polycrystalline. This material can also be divided into two families: polycrystalline diamond (PCD), and polycrystalline cubic boron nitride (PCBN or CBN).

In addition to the above materials, new cutting materials and coatings are being introduced regularly. For engine lathe operations, it is unlikely that cutting materials more exotic than tungsten carbide will ever be needed; but as your career advances, chances are good that you will someday encounter machining challenges that will mandate the use of cermet, ceramic, diamond, or other exotic cutting materials.

Drill Bits

Drill bits are the most universal cutting tools used in any shop, and the vast majority of twist drills in use today are of the following types: Screw Machine Length (a short length range), Jobber Length (medium length), or Taper Length (long length). For example, general purpose $3/8$" drills of these three types have the following length dimensions. Screw Machine Length: 1

$7/8$" flute, 3 $1/4$" overall. Jobber Length: 3 $5/8$" flute, 5" overall. Taper Length: 4 $1/4$" flute, 6 $3/4$" overall.

In addition, diameter sizes for inch size drills are grouped into three different size codes.

1. Number sizes, which range from #97 (.0059" diameter) to #1 (.2280" diameter).
2. Fractional sizes, ranging from $1/64$" (.0156" diameter) to 1" in diameter (although taper length drills are available in diameters up to 1 $3/4$").
3. Letter sizes that range from "A" (.2340" diameter) to "Z" (.4130" diameter).

Diameters of metric drills are always expressed in millimeters.

Screw Machine Length drills (also known as stub drills) were developed for use in screw machines where their short flutes and short overall length provide maximum rigidity. They are available in fractional sizes from $3/64$" to 1", number sizes from #60 through #1, every letter size, and most metric sizes from 1 mm to 25 mm. Their lengths range from 1 $3/8$" (35 mm) to 6" (152 mm).

Jobber Length drills are available in every fractional size up to $11/16$", every number size, every letter size, and metric sizes through 17.50 mm. Lengths range from $3/4$" (19 mm) to 7 $5/8$". They perform well in many high production operations. They have straight shanks and short flutes, and they are comparatively rigid due to their short overall length-to-diameter ratio.

Taper Length drills are popular for use in lathes. They are available in the same sizes as screw machine length drills (see above), and in lengths from 2 $1/4$" (57 mm) to 11" (279 mm).

98   Lathe Operation and Maintenance

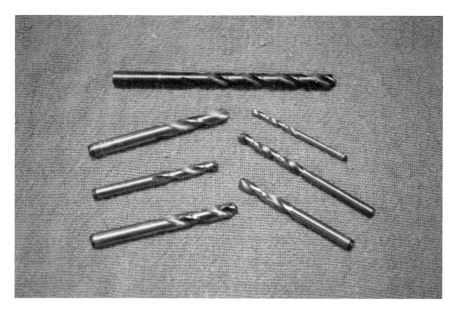

Figure 5.4 Screw machine length drill bits—the longer drill at the top is a jobber length drill and is shown for comparison purposes.

Figure 5.5 A Morse taper shank drill bit.

Drills with Morse taper shanks are available for a wide range of sizes—from $3/8$" to $3\ 1/2$". Due to their more robust shanks, these drills will provide more rigidity than straight shank twist drills.

For some deep-hole drilling operations, *aircraft extension* drills can be used. These drills were designed for the aircraft industry, primarily for drilling rivet holes into airframes. They have a long shank and a relatively short flute area. They are available in 12" and 18" lengths, and in the following diameters: fraction sizes from $3/64$" to $1/2$"; and wire gage sizes from #1 through #60. **Note:**

Because of the length of the shank, you should center drill the hole location before you use an aircraft drill.

When very deep holes (in excess of 8" and up to 14") must be drilled, use an *extra long* drill bit. These drill bits differ from aircraft drills because they have an extra long flute length. For example, 8" long drills will have a flute length of approximately 5 $1/2$"; 12" long drills have a 9" flute length; and 18" long drills have a 14" flute length. Extra long drills are available in fraction sizes.

Center Drills

Center drills are used to precisely locate a hole to the workpiece. They have a large shank diameter as compared to the actual drill end. This increase in diameter provides the center drill with extra rigidity. Center drills come in several sizes and lengths. Sizes typically range from #000 (.020" drill diameter) to #10 ($3/8$" drill diameter). Standard lengths range from 1 $1/4$" (#000) to 3 $3/4$" (#10). For sizes #1 to #8, extra long center drills are available in 3", 4", 5", and 6" lengths.

The majority of commercially available center drills have a 60°

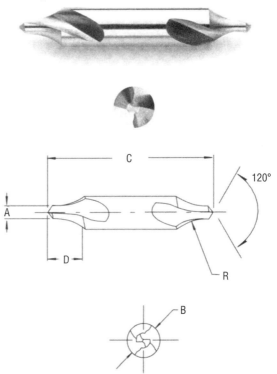

Figure 5.6 *A typical radius center drill. (Courtesy and copyright KEO Cutters, Division of Masco Tech.)*

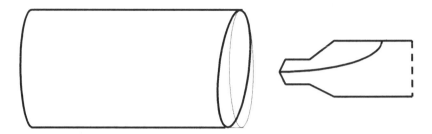

Figure 5.7 This part was cut with a slight angle. A center drill must be used to make certain that the hole is centered in the workpiece.

straight taper. This is the taper found on the tips of most lathe centers. There is also a *radius* type, which has no pilot-drill end but cuts a continuous radius and is also compatible with a 60° lathe center. These types of center drills allow for increased spindle speeds and reduce the possibility of pilot drill breakage in the hole.

The use of center drills will also allow you to start a pilot hole in a workpiece that does not have the work surface parallel to the machine spindle, as would be the case with some castings, or when the workpiece was cut at an angle, as shown in *Figure 5.7*.

Taps and Reamers

Taps are tools used to cut internal threads in a workpiece. Like drills, they come in an almost infinite number of styles and sizes. On the lathe, they are used in much the same manner as drills, although taps usually operate at lower speeds than drills. When using a tap to cut threads, it is necessary to drill a pilot hole to a specified size for the thread. Lists of recommended predrilled hole sizes for taps can be found in most handbooks for machinists.

Reamers are multifluted cutting tools used to machine an existing hole to a final size and improve surface finish. They are sometimes used to enlarge a drilled hole to a precise size before tapping. Like taps, they are typically made from solid high-speed steel blanks, but solid carbide or carbide-tipped reamers are available for heavy-duty applications. Reamers should not be used to remove large amounts of stock—they are designed to cut away small amounts of material.

Boring Tools

Single-point boring tools

Boring consists of enlarging an existing hole to a specified dimension with a boring tool. To allow for the tool to enter the workpiece, the existing hole must be large enough to accept the tool. Because boring is an internal machining operation, great care must be taken to assure that chips are evacuated from the tool bit. Otherwise, it is quite easy to recut the chips and damage the surface finish of the part.

Single-point boring tools use a replaceable brazed point high-speed steel or carbide-tipped tool bit, and come in a wide range of shank diameters, lengths, and minimum bore sizes as shown in *Tables 5.10* through *5.15*. Both tip materials allow for customization for special applications. For best results, keep the length of the boring tool as short as possible, as this will increase rigidity.

Indexable insert boring tools

Boring bars that use indexable inserts come in a vast array of dimensions—some are very small in size—allowing their use in smaller diameter preexisting holes than can be entered by a single-point tool. For the majority of work, either diamond or triangular shaped inserts can be used.

Table 5.10: Style TSA Brazed Single-point Boring Tools for 90° Boring Bar. *(Source, Kennametal Inc.)*

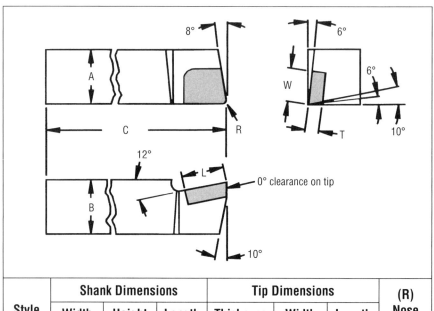

Style	Shank Dimensions			Tip Dimensions			(R) Nose Radius
	Width (A)	Height (B)	Length (C)	Thickness (T)	Width (W)	Length (L)	
TSA 5	5/16	5/16	1 1/2	3/32	3/16	5/16	1/64
TSA 6	3/8	3/8	1 3/4	3/32	3/16	5/16	1/64
TSA 8	1/2	1/2	2 1/4	1/8	5/16	7/16	1/32

Table 5.11: Style TSC Brazed Single-point Boring Tools for 60° Boring Bar. *(Source, Kennametal Inc.)*

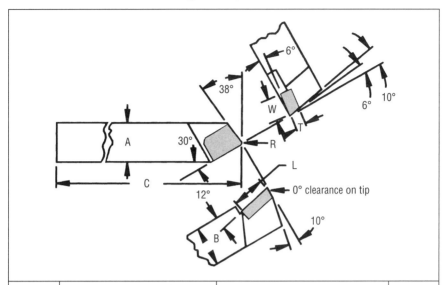

Style	Shank Dimensions			Tip Dimensions			(R) Nose Radius
	Width (A)	Height (B)	Length (C)	Thickness (T)	Width (W)	Length (L)	
TSC 5	5/16	5/16	1 1/2	3/32	3/16	5/16	1/64
TSC 6	3/8	3/8	1 3/4	3/32	3/16	5/16	1/64
TSC 8	1/2	1/2	2 1/4	1/8	5/16	7/16	1/32
TSC 10	5/8	5/8	3	5/32	3/8	9/16	1/32
TSC 12	3/4	3/4	3 1/2	3/16	7/16	5/8	1/32

Table 5.12: Style TSE Brazed Single-point Boring Tools for 45° Boring Bar. *(Source, Kennametal Inc.)*

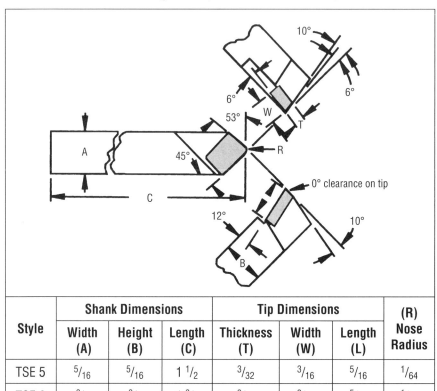

Style	Shank Dimensions			Tip Dimensions			(R) Nose Radius
	Width (A)	Height (B)	Length (C)	Thickness (T)	Width (W)	Length (L)	
TSE 5	$5/16$	$5/16$	$1\ 1/2$	$3/32$	$3/16$	$5/16$	$1/64$
TSE 6	$3/8$	$3/8$	$1\ 3/4$	$3/32$	$3/16$	$5/16$	$1/64$
TSE 8	$1/2$	$1/2$	$2\ 1/2$	$1/8$	$5/16$	$7/16$	$1/32$
TSE 10	$5/8$	$5/8$	3	$5/32$	$3/8$	$9/16$	$1/32$
TSE 12	$3/4$	$3/4$	$3\ 1/2$	$3/16$	$7/16$	$5/8$	$1/32$

Table 5.13: Style TRC Brazed Single-point Boring Tools for 60° Boring Bar. *(Source, Kennametal Inc.)*

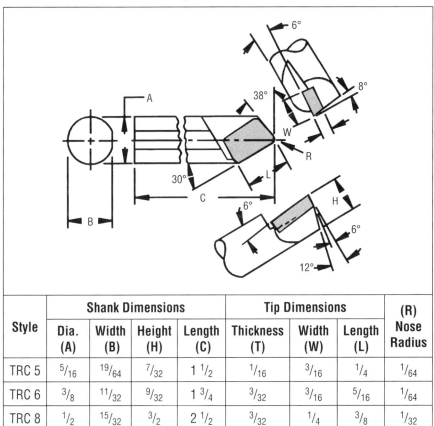

Style	Shank Dimensions				Tip Dimensions			(R) Nose Radius
	Dia. (A)	Width (B)	Height (H)	Length (C)	Thickness (T)	Width (W)	Length (L)	
TRC 5	5/16	19/64	7/32	1 1/2	1/16	3/16	1/4	1/64
TRC 6	3/8	11/32	9/32	1 3/4	3/32	3/16	5/16	1/64
TRC 8	1/2	15/32	3/2	2 1/2	3/32	1/4	3/8	1/32

Table 5.14: Style TRE Brazed Single-point Boring Tools for 45° Boring Bar. *(Source, Kennametal Inc.)*

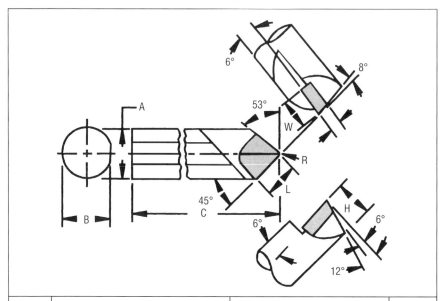

Style	Shank Dimensions				Tip Dimensions			(R) Nose Radius
	Dia. (A)	Width (B)	Height (H)	Length (C)	Thickness (T)	Width (W)	Length (L)	
TRE 5	5/16	19/64	7/32	1 1/2	1/16	3/16	1/4	1/64
TRE 6	3/8	11/32	9/32	1 3/4	1/16	3/16	1/4	1/64
TRE 8	1/2	15/32	3/2	2 1/2	3/32	5/16	3/8	1/32

Table 5.15: Style TRG Brazed Square End Boring Tools.

(Source, Kennametal Inc.)

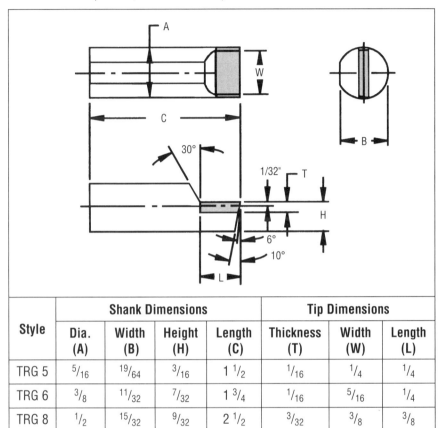

Style	Shank Dimensions				Tip Dimensions		
	Dia. (A)	Width (B)	Height (H)	Length (C)	Thickness (T)	Width (W)	Length (L)
TRG 5	$5/16$	$19/64$	$3/16$	$1\,1/2$	$1/16$	$1/4$	$1/4$
TRG 6	$3/8$	$11/32$	$7/32$	$1\,3/4$	$1/16$	$5/16$	$1/4$
TRG 8	$1/2$	$15/32$	$9/32$	$2\,1/2$	$3/32$	$3/8$	$3/8$

Radius Cutters

Occasionally you may have to cut a round or spherical shape onto the workpiece. You might think that this can't be done; there is, however, a tool specifically designed for the purpose—the radius cutter. It mounts in the compound's T-slot, and there are several sizes available to suit the size of the radius and lathe capacity.

When discussing a radius cutter, we also need to think about the terminology used to describe the shape. We use *convex* to describe an outward protruding radius or sphere. *Concave* describes the radius cut into the surface of the workpiece. Think of the convex shape as the *ball*, the concave shape as the *socket*.

It should become obvious that the radius cutter must be set up differently for each shape. What's interesting is that in order to cut a convex shape, the cutter must be set in a concave configuration, and cutting a concave shape requires a convex arrangement. This will be readily apparent when you use a radius cutter.

An alternative to using a radius cutter to machine a spherical shape would be to grind a tool bit with the appropriate radius. For speed and production, this method would deserve closer attention. There are limits as to just how much radius you can cut. Typically, it is dependent upon the size of the tool bit blank.

Form Tools

Forming is another simple machining process where a cutting tool is plunged into a workpiece to create a shape or form. There are few limits to the shapes that can be cut into the workpiece with form tools. The tools themselves must be machined to very precise dimensions as they must possess the rake and clearance specific to the material and shape to be machined.

Form tools allow the machine operator to cut complicated patterns without the need for a series of special tools. Cutters with concave and convex shapes can easily machine a desired radius very quickly and efficiently.

The design and application of form tools must be carefully considered in the planning stages of the job. Don't expect the form tool to remove large amounts of material (sometimes called "hogging") all in one setting. Form tools are best utilized during the final sizing of the workpiece. In other words, hog out the majority of the material and leave .100" (2.50 mm), or less, for the final cut.

Knurling Tools

Knurling tools are used to produce patterns by pushing a pair of sharp rollers into the workpiece as it rotates. Although knurling tools may remove some material, their primary function is to redistribute it into functional (or

Figure 5.8 *A knurling tool.*

decorative) surface textures. These textures are usually diamond-shaped, although variations are also possible. One variation, parallel lines, is typically used to marginally increase the diameter of a workpiece for tight assembly with another part (as in making a shaft slightly oversize so it can be inserted into a hole for a "press fit").

To set up a knurling tool, the tool is mounted in the normal toolholder, aligned perpendicular to the workpiece.

Tool Post Grinders

A tool post grinder can broaden the capabilities of your lathe by allowing you to grind the exterior, interior, and surface finishes of a workpiece. It mounts on the cross-slide and replaces the standard tool mount. This can be a very useful accessory for some machinists, as it can be used to produce very fine surface finishes—much smoother than can be achieved with a tool bit.

Figure 5.9 *A tool post grinder.*

When grinding, especially with very abrasive wheels, a good deal of grit is produced. It comes not only from the workpiece, in the form of very small metallic flakes, but also from the wheel itself as it sheds particles from its surface. These small particles can (and probably will) find their way onto the sliding surfaces of the saddle, cross-slide, and compound-rest, and they can be devastating to your lathe. Extreme care must be taken to keep this from happening—lathes are not intended for such extreme service, and they can be ruined in very short order if not treated properly.

Care of Cutting Tools

Care of your cutting tools is as important as their selection and purchase. Tool bits should not be carelessly pitched into a drawer, nor should they be stored where they can become rusty. Their sharp edges must be protected so they do not become nicked—small craters in the cutting edge will cause the tool to run hot and also do damage to the surface finish of the workpiece.

Store your tools in racks or drawer assemblies designed for this purpose. Racks can be made by drilling appropriate size holes in a wooden block, which will also allow you to organize your tools by size or type. If tool bits must be stored in a steel cabinet or drawer, wrap a piece of masking tape around each bit or place a plastic sleeve over the cutting edges.

Review Questions for Chapter 5

1. HSS stands for:

    A. High Spindle Speed

    B. Hot Series Steel

    C. High Speed Spindle

    D. High Speed Steel.

2. A brazed single-point cutting tool can cut _____ times faster than a high speed tool bit.

    A. 1

    B. 2

    C. 3

    D. 4.

3. Tool coating such as TiN, CVD, and PVD are used to increase _____.

    A. Tool life

    B. Abrasion resistance

    C. Productivity

    D. All of the above.

4. A radius center drill is used on parts that may be cut at an angle.

   A. True

   B. False.

5. Knurling tools usually impart a _____ texture onto the surface of a part.

   A. Round

   B. Diamond

   C. Square

   D. Pear-shape.

Workholding Devices

In this chapter a variety of workholding devices and tools will be described. Choosing the best workholding apparatus for a particular job is essential for maximizing efficiency and productivity. Try to become familiar with how these mechanisms work, and experiment with as many as possible so you can gain first-hand experience of the advantages and shortcomings of each device. First, we will look at the nose of the spindle so that you can identify the type of nose on your lathe, and find out how to mount and dismount chucks from your spindle's nose. Then we will examine the different types of chucks and collets available for your lathe.

Spindle Noses

There are four basic types of spindle noses found on lathes: threaded, short taper, cam lock, and long taper key drive. Each one performs the same job—it mounts the chuck or faceplate to the spindle—but each accomplishes its task with a different method of mechanical attachment. The type of nose on your lathe was determined when it was manufactured, and it cannot be reconfigured by the operator once the lathe has been installed. Each type of mount has particular advantages, and disadvantages, as discussed below.

Threaded spindle nose

The threaded spindle nose (pictured in Chapter 3, *Figure 3.4*) is typically found on small machines with a swing of less than 14". This mount makes it quite simple for the machinist to install and remove the chuck. The threaded mounting system, however, has some inherent disadvantages. If the machine is equipped with a sufficiently powerful motor, removal of the chuck or backing plate can become somewhat troublesome, because the chuck is increasingly tightened as the machine expends large amounts of energy to turn the workpiece. This means that unscrewing the chuck from the spindle can sometimes be very difficult, especially after machining a heavy workpiece at high speed. Another thing to keep in mind when using a threaded chuck is that reversing the spindle, especially when it is rapidly accelerated, can actually unscrew the chuck from the spindle and send it flying around the shop.

The majority of thread-mounted chucks will use a transition or converter plate onto which the chuck is bolted. Alignment is accomplished by machining a small step onto the transition plate that matches a recess on the backside of the chuck. Should the transition get knocked out of alignment, simply remove the chuck and realign it on the plate. It is best to buy the backing plate when you purchase the chuck, so that the two are ideally mated.

Threaded spindle chuck installation procedure.

1. With the lathe Off, lock the spindle so that it does not rotate. This can be done by putting the machine into back-gear, or into a low speed gear.

2. Place a wooden block or plank onto the bed of the machine under the spindle.

3. Clean the spindle and chuck threads. Be sure to remove any debris that may cause the assembly to misthread.

NEVER switch the lathe from forward to reverse in rapid succession when a threaded type chuck is installed. Rapid reversing of the motor may cause the chuck to become loose and unwind itself from the spindle, which can cause great bodily injury. Always have your instructor or supervisor check your chuck installation procedure before you start your machine.

4. Place the chuck onto the spindle. Rotate the chuck in a clockwise (CW) rotation slowly until the chuck just touches the spindle.

5. Rotate the chuck 45°– 60° (degrees) counterclockwise (CCW), then quickly rotate the chuck clockwise (CW) to seat the chuck against the spindle. Do not try to overpower the chuck when you seat it against the spindle. This will cause the chuck to lock into place and make its eventual removal very difficult.

6. Remove the wooden block/plank.

Threaded spindle chuck removal procedure.

1. With the lathe Off, lock the spindle so that it does not rotate. This can be done by putting the machine into back-gear, or into a low speed gear.

2. Place a wooden block or plank onto the bed of the machine under the chuck.

3. Place the chuck key into the chuck. The chuck key location should be slightly behind the centerline of the spindle.

USE extreme caution when removing any chuck from the lathe!

4. With a quick and decisive jerk, unlock the chuck from the spindle. **Note:** If this procedure fails to loosen the chuck for removal, use the following procedure:

   A. Place a large adjustable wrench onto one of the chuck jaws.

   B. Ease the pressure onto the wrench until the chuck releases from the spindle. Be sure that your hands are out of harm's way in case the wrench slips off.

5. Rotate CCW until the chuck releases from the spindle. Be sure to keep your hands and fingers from underneath the chuck. It should drop, ever so slightly, onto the wooden plank.

6. If the chuck is heavy, get help when removing it from the machine. Store the chuck in a safe place.

NEVER stand in front of the chuck when you initially start the lathe after the chuck has just been installed. Stand back! Once you or your instructor/supervisor is satisfied that the chuck will not come off, then take your normal position or stance in front of the machine.

Table 6.1: Threaded Spindle Nose Dimensions.

(Courtesy of Toolmex Corporation.)

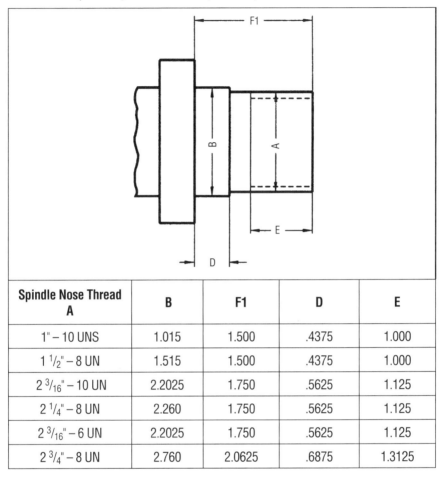

Spindle Nose Thread A	B	F1	D	E
1" – 10 UNS	1.015	1.500	.4375	1.000
1 1/2" – 8 UN	1.515	1.500	.4375	1.000
2 3/16" – 10 UN	2.2025	1.750	.5625	1.125
2 1/4" – 8 UN	2.260	1.750	.5625	1.125
2 3/16" – 6 UN	2.2025	1.750	.5625	1.125
2 3/4" – 8 UN	2.760	2.0625	.6875	1.3125

All dimensions in inches.

Short taper (American Standard) spindle nose

On machines fitted with the American Taper A-type spindle nose (pictured in Chapter 3, *Figure 3.5*), the procedure for installation and removal of the chuck is very straightforward. The chuck is held in position with standard socket-head fasteners and is aligned to the spindle nose with a short tapered snout. This type of chuck attachment is typically used on machines with higher power outputs, and where reversing of the spindle is to be expected.

There are two classes of short taper nose: A-1 and A-2. The A-1 has one bolt circle, and the A-2 has two. Chucks with A-2 mounts may be installed on both A-1 and A-2 noses, but chucks with A-1 mounts can only be installed on A-1 noses.

Table 6.2: Short Taper Spindle Nose Dimensions.
(Courtesy of Toolmex Corporation.)

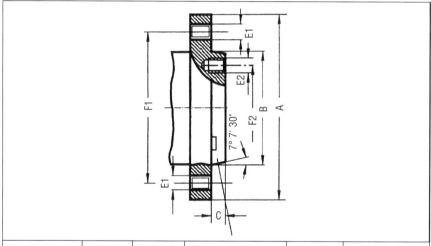

Spindle Nose Designation	F1	F2	B (tolerance)	C (max.)	Thread Size E1 & E2
A-4	3.2500	–	2.5005 (+.0005)	$7/16$	$7/16$ – 14 UNC
A-5	4.1250	2.4374	3.2505 (+.0005)	.5625	$7/16$ – 14 UNC
A-6	5.2500	3.2500	4.1880 (+.0005)	.6250	$1/2$ – 13 UNC
A-8	6.7500	4.3750	5.50075 (+.0005)	.6875	$5/8$ 11 UNC
A-11	9.2500	6.5000	7.75075 (+.0005)	.7500	$3/4$ – 10 UNC
A-15	13.000	19.750	11.251 (+.001)	.8125	$7/8$ – 9 UNC
A-20	18.250	14.500	16.251 (+.001)	.8750	1 – 8 UNC

All dimensions in inches.

There will be a minimum of three fasteners on the mounting. On larger machines, you may find more fasteners to attach the chuck to the spindle. These chucks are very heavy. Use your best judgment when working with heavy parts—get help when you need it, not after!

Short taper chuck installation procedure.

1. With the lathe Off, lock the spindle so that it does not rotate. This can be done by putting the machine into back-gear, or into a low speed gear.
2. Place a wooden block or plank onto the bed of the machine under the spindle.
3. Ensure that there is no debris in the chuck, the spindle nose, or the fasteners.
4. Place the chuck onto the spindle, and align the bolt holes so that the fasteners will enter their holes on the spindle.
5. Tighten the fasteners in the recommended sequence. They should be tightened to a specified torque, so check your handbooks for additional information.
6. Remove the wooden plank.

Short taper chuck removal procedure.

1. With the lathe Off, lock the spindle so that it does not rotate. This can be done by putting the machine into back-gear, or into a low speed gear.
2. Place a wooden block or plank onto the bed of the machine under the chuck.
3. Loosen the first fastener with the appropriate wrench or chuck key.
4. Rotate the chuck to the next fastener position and repeat the removal procedure. Repeat the entire process until the chuck is loose.

5. Pull the chuck away from the spindle onto the wooden plank.

6. If the chuck is heavy, get help when removing it from the machine. Store the chuck in a safe place.

Cam-lock spindle nose

On machines fitted with D-type cam-lock spindle noses (pictured in Chapter 3, *Figure 3.6*), the procedure for installation and removal of the

Table 6.3: Cam-lock Spindle Nose Dimensions.
(Courtesy of Toolmex Corporation.)

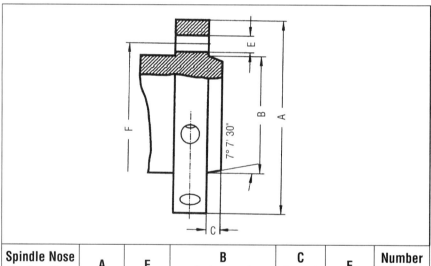

Spindle Nose Designation	A	F	B (tolerance)	C (max.)	E	Number Of Holes
D-1	3.622	2.7820	2.1250 (+.00025)	7/16	.5937	3
D-4	4.606	3.2500	2.5005 (+.0005)	7/16	.6562	3
D-5	5.748	4.1250	3.2505 (+.0005)	1/2	.8750	6
D-6	7.126	5.2500	4.1880 (+.0005)	9/16	1.000	6
D-8	8.858	6.7500	5.50075 (+.0005)	5/8	1.125	6
D-11	11.732	9.5250	7.75075 (+.0005)	11/16	1.250	6
D-15	15.866	13.000	11.251 (+.001)	3/4	1.375	6

All dimensions in inches.

chuck is straightforward. The chuck is held in position with cam-locks around its perimeter, and is aligned to the spindle with a short tapered snout. This type of chuck attachment is typically used on machines with higher power outputs, and where reversing of the spindle is to be expected.

There will be a minimum of three cam-locks on the mounting. On larger machines, you may find as many as six or more cam-locks to attach the chuck to the spindle. These chucks are very heavy. Use your best judgment when working with heavy parts—get help when you need it, not after!

Cam-lock chuck installation procedure.

1. With the lathe Off, lock the spindle so that it does not rotate. This can be done by putting the machine into back-gear, or into a low speed gear.
2. Place a wooden block or plank onto the bed of the machine under the spindle.
3. Ensure that there is no debris in the cam-lock holes or on the tapered snout areas on both the machine and chuck.
4. Place the chuck onto the spindle, and align the cam-locks so that the dowels on the chuck will enter their holes on the spindle.
5. Tighten the cam-locks in the recommended sequence.
6. Remove the wooden plank.

Cam-lock chuck removal procedure.

1. With the lathe Off, lock the spindle so that it does not rotate. This can be done by putting the machine into back-gear, or into a low speed gear.
2. Place a wooden block or plank onto the bed of the machine under the chuck.

3. Loosen the first cam-lock with the appropriate key.

4. Rotate the chuck to the next cam-lock position and repeat the loosening procedure. Repeat the process until the chuck is free.

5. Pull the chuck away from the spindle onto the wooden plank.

6. If the chuck is heavy, get help when removing it from the machine. Store the chuck in a safe place.

Long taper key drive nose

The long taper key, L-Mount nose (pictured in Chapter 3, *Figure 3.7*) is reserved for robust machines that are often engaged in heavy machining.

Table 6.4: Long Taper Key Spindle Nose Dimensions.
(Courtesy of Toolmex Corporation.)

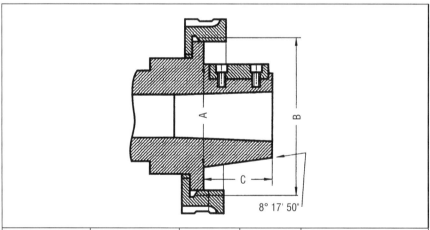

Spindle Nose Designation	Thread B	C	A	Key
L00	3 3/4" – 6	2	2.750	3/8 × 3/8 × 1 1/2
L0	4 1/2" – 6	2 3/8	3.250	3/8 × 3/8 × 1 3/4
L1	6" – 6	2 7/8	4.125	3/8 × 5/8 × 2 3/8
L2	7 3/4" – 5	3 3/8	5.250	3/4 × 3/4 × 2 7/8
L3	10 3/8" – 4	3 7/8	6.500	1 × 1 × 3 1/4

Because these chucks are large, they are exceptionally heavy. The tapered snout perfectly aligns the chuck to the spindle every time, just like the cam-lock types. This is a real advantage over threaded spindle machines that put you at the mercy of the accuracy of the backing plate to ensure that the chuck aligns itself properly—even if the threads are precisely cut, the slack between the threads can be enough to cause a small misalignment.

Long taper key chuck installation procedure.

1. With the lathe Off, lock the spindle so that it does not rotate. This can be done by putting the machine into back-gear, or into a low speed gear.
2. Place a wooden block or plank onto the bed of the machine under the spindle.
3. Slide the chuck onto the spindle. Be sure that the key is aligned with the chuck keyway.
4. Tighten the lock ring by hand until the chuck is tight.
5. Use the appropriate size wrench to tighten the lock ring on the spindle.
6. Remove the wooden plank.

After the chuck is installed onto the spindle, clamp a piece of round stock in the jaws. Position a test indicator so that the tip contacts the round stock. Rotate the spindle (chuck) by hand and check the eccentricity of the shaft. If possible, adjust the chuck until the eccentricity is as close to ZERO as possible. This test will not be required if you have installed a 4-jaw independent chuck.

Long taper key chuck removal procedure.

1. With the lathe Off, lock the spindle so that it does not rotate. This can be done by putting the machine into back-gear, or into a low speed gear.

2. Place a wooden block or plank onto the bed of the machine under the chuck.

3. Use the appropriate size wrench to loosen the lock ring on the spindle.

4. Loosen the lock ring until the chuck is loose from the spindle.

5. Get help, as needed, to remove the chuck from the machine.

Face Plate

The face plate (or "gap" face plate) is an extremely handy means to mount certain workpieces, especially those with flat bottoms that can be bolted directly to the plate and then mounted to the spindle through the procedures mentioned above. A series of slots that are cast or machined into the face plate allow for the mounting of odd shaped workpieces for turning, boring, or facing operations.

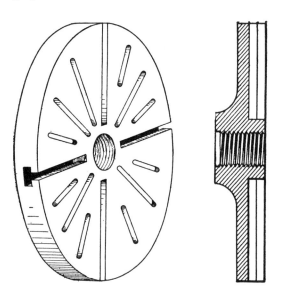

Figure 6.1 A face plate is the most basic means used to mount a workpiece to the lathe spindle. This face plate for a threaded spindle nose is large enough to accommodate a variety of workpieces.

Some shops will customize a face plate by drilling mounting holes directly into the plate to facilitate mounting of unusual size workpieces. If spaced properly, it is possible to drill a good number of holes into the face plate without weakening it, but it should be kept in mind that an equal (or almost equal) amount of material should always be removed from opposite sides of the plate's surface so that it does not become unbalanced and cause vibrations when rotated.

The face plate is also used when turning parts between centers, as you will discover in the next section.

Mounting Work Between Centers

The majority of lathe spindles have a Morse taper machined directly into the spindle. This is to allow for the installation of a "center" so that workpieces can be mounted "between centers"—the second center being installed in the tail stock. Centers are normally classified as either "live" or "dead." Traditionally, centers used in the head stock spindle were referred to as live, and those used in the tail stock were called dead, primarily because the head stock rotates and the tail stock does not. A more accurate description of a live center is one that has bearings installed within its head to allow the pointed end of the spindle to revolve while the shaft is stationary. The pointed end on a center is machined and ground at 60°, and the other end is machined with a Morse taper. A live center is shown in *Figure 6.2*. Dead centers, on the other hand, are solid throughout (*Figure 6.3*).

A series of short adapters may be required to install the desired size center: e.g., a #5 spindle with a #3 center. Most machine manufacturers supply adapters that will allow the installation of the same size centers in both the spindle taper and in the tail stock, making it possible to use equal size centers on both ends of the workpiece.

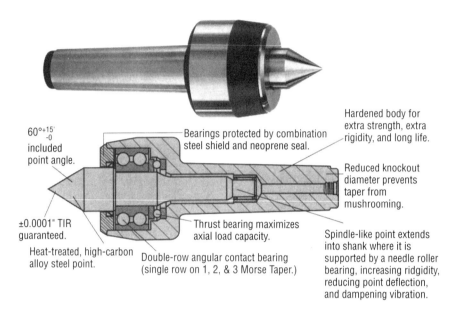

Figure 6.2 A live center. (Courtesy, Royal Products.)

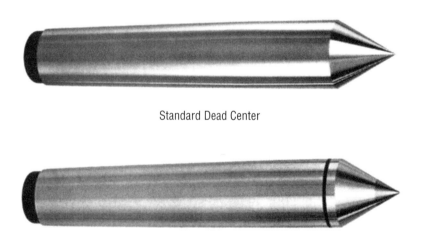

Figure 6.3 Dead centers. (Courtesy, Royal Products.)

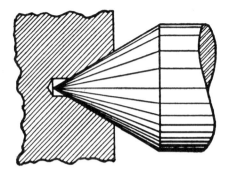

Figure 6.4 A center properly seated into a workpiece. The pointed nose of the center rests in the drilled pilot hole.

Machining a workpiece between centers is, in the opinion of most machinists, the most fundamental of turning procedures, but first the workpiece must be center-drilled on each end so that it can be placed between the center in the spindle and one located in the tail stock. If a dead center is used in the tail stock, it will require an extreme pressure (EP) lubricant to ensure proper operation and extend the useful life of the center. Live centers do not require lubricant because the point of the center rotates with the workpiece.

To provide the motive force required to revolve the workpiece, a small face plate must first be mounted onto the spindle. Then, a *drive-dog* (*Figure 6.5*) is slipped over and then clamped to the workpiece. The "*tail*" of the dog is positioned so that it fits through one of the slots in the face plate. Once everything is properly positioned, the spindle will be able to drive the face plate, which will drive the dog, which will in turn rotate the workpiece (*Figure 6.6*).

Lathe dogs come in many configurations to accommodate most operations, but sometimes it may be necessary to customize an existing dog to conform to the requirements dictated by a difficult size workpiece.

Once the workpiece is positioned between centers, slack is then taken up with the tail stock adjusting wheel. It is not necessary to get it as tight as possible, just to impose a sufficient thrust load to keep the workpiece from moving during machining. Although a thrust load of two-thirds of the workpiece weight is considered sufficient, in most instances the workpiece should be tightened just enough so that it will revolve easily by hand, but without any wobble.

The majority of lathe centers are made from HSS. However, for those applications where heavy wear is expected, carbide tipped versions are available. They are sometimes necessary for machining exotic materials and composites that are exceptionally abrasive and can rapidly wear out a standard HSS center.

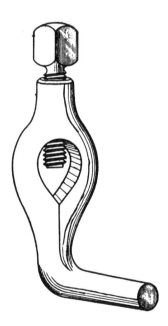

Figure 6.5 *A lathe dog must be attached to the workpiece in order for it to be driven for machining.*

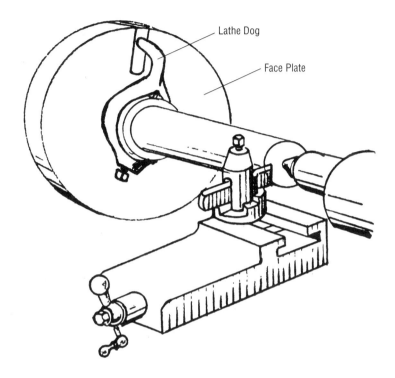

Figure 6.6 *In this illustration, the lathe dog is secured to the workpiece, and the tail of the dog is driven by the face plate.*

Other lathe centers

Some type of lathe center is necessary when machining long parts and removing lots of material during roughing cuts. The Morse taper found in the tail stock permits the mounting of several types of lathe centers, including the *bull-head* center. The bull-head (*Figure 6.7*) is much larger than standard centers, and is used to center and support hollow pipes and tubes that have openings too large for standard 60° centers. The angle of the bull-head center is 70°, or 75°, and its diameter is determined by the size of its Morse taper mount. Commercially available sizes for bull-head dead centers range in overall diameter from 2 $1/8$" to 3 $5/8$", with point diameters ranging from $^1/_2$" to 1". Live centers range in overall diameter from 6 $^1/_4$" to 12 $^1/_4$", with point diameters ranging from 2 $^3/_4$" to 5 $^3/_4$".

Special lathe centers are available with interchangeable points. These centers can be handy when machining certain small diameter stock. In

Figure 6.7 *A bull-head center.*

Figure 6.8 *This pipe is supported at the tail stock end with a bull-head center.*

addition to standard 60° points in various diameters, centers with concave (female point) ends are also included. These can be used to machine parts that cannot be center drilled. The reverse point acts as a cup to cradle the workpiece.

Chuck Designs

The most common method used to hold and position parts on the engine lathe is the chuck. Chucks are available in a variety of sizes and jaw configurations. Sizes range from 3", to over 32", and chucks may have from two to six jaws.

The majority of chucks (2-, 3-, and 6-jaw types) have *universal* jaws (*Figure 6.9*). This means that the jaws are synchronized to open and close together when they are tightened or loosened. Four-jaw chucks, on the other

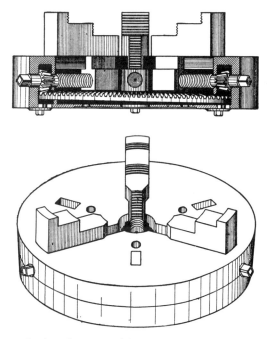

Figure 6.9 *A 3-jaw chuck with universal jaws.*

hand, are said to have *independent* jaws because each jaw can be positioned individually.

Chuck jaws may have machined surfaces on both interior and exterior faces in order to increase their holding capability. In addition, some chuck jaws may have the ability to be changed from external holding to internal holding positions. This is done by repositioning the jaws in the opposite direction.

Most chuck jaws are opened and closed with a *chuck key* that has a square or hexagon end. Chucks that have pneumatic actuation to open and close the jaws are known as power chucks, and they are especially useful when large production runs are expected.

The jaws in the chuck are actuated with an internal scroll gear when the key is turned. These gears are sometimes subject to chip intrusion because of their location and lack of protection. Take great care when repositioning the jaws to keep errant chips out of the gears. Loose chips in the wrong place can cause an abrasive cutting action to occur that will shorten the life expectancy of your workholding tools.

The method used to attach the chuck to the lathe spindle will influence whether or not the chuck will *run true*. For accurate machining, the chuck must run true and concentric to the center of the spindle rotational axis, and, to aid in chuck alignment, some chuck manufacturers install special adjustment screws that allow the machinist to align, or "dial in," the chuck to the spindle. When working with exacting specifications, this is the best type of chuck to use. Also, keep in mind that the chuck can be knocked out of alignment by crashing the carriage or cross-slide into the spindle while the machine is running. Be careful not to make this mistake!

While not as fundamental to a chuck's operation as the configuration of the jaws, the diameter of the center- (or through-) hole in the chuck can determine its usefulness. If you work on large diameter parts and need a certain amount of depth for clamping, a large center-hole may make the job a lot easier to set up. Also, take into consideration the maximum diameter that will pass easily through the spindle. You may find that some chucks have a larger center-hole than the spindle. This may be a moot point, but you need to ponder the advantages and disadvantages of both components.

Although every chuck has its own operational limits and balance requirements, *Tables 6.5* through *6.7* provide guidelines that may be useful in selecting and using chucks on your lathe.

Table 6.5: Chuck Gripping Power.*

Chuck Size	Semi-Steel Body	Precision Chucks
3 1/4"	2200	2200
4"	3750	3750
5"	5300	5300
6 1/4"	6800	6800
8"	8200	8200
10"	10000	10000
12 1/2"	12000	12000
15 3/4"	14300	14300
20"	16000	–
25"	17700	–
31 1/2"	19800	–

* Approximate pressure in pounds per square inch.

Table 6.6: Recommended Maximum RPM.

Chuck Size	Semi-Steel Body	Precision Chucks
3 1/4"	4000	6000
4"	3500	5200
5"	3200	4800
6 1/4"	3000	4500
8"	2500	4000
10"	2000	3500
12 1/2"	1500	2800
15 3/4"	1000	2000
20"	700	1200
25"	500	1000
31 1/2"	300	–

Table 6.7: Recommended Balancing Limits.

Chuck Size	Balance (ounce-inch)
3 1/4"	.15
4"	.22
5"	.32
6 1/4"	.44
8"	.62
10"	.87
12 1/2"	1.25
15 3/4"	1.94

2-jaw chucks

This is somewhat of a specialty chuck and may not be found in all machine shops. However, when you have a part that is not quite round, and you need to perform several turning, boring, or facing operations, the 2-jaw chuck may prove to be the best selection.

The 2-jaw chuck will normally use "top," rather than standard, jaws. Top jaws are much wider and thicker than standard or reversible jaws, so each jaw has a maximized gripping area. They can be purchased in cast iron, steel, or aluminum—the actual material you use will be dependent upon the shape and size of the workpiece.

For really odd-shaped workpieces, top jaws can be mounted onto the chuck and machined to any desired shape and/or contour with a milling machine to provide a custom workholding tool. Even though the custom jaws can only be used with a specific size workpiece, they will greatly reduce setup times if a large production run is planned.

Two-jaw chucks are normally available in the following sizes: 4", 6", 8", 10", and 12 $^1/_2$".

3-jaw universal chuck

This is most machinists' favorite chuck because it is easy to use and can hold any round part with a certain degree of accuracy. Each jaw moves simultaneously as it is opened and closed, so with very little effort it is possible to accurately center the workpiece as the jaws do most of the work for you.

Three-jaw chucks usually come with standard jaws, reversible master jaws, or top jaws. If your chuck is the standard type, it will have a set of external and internal jaws. The jaws cannot be interchanged with one another—they are numbered and must be installed in the correct order for the workpiece to run true. Look at the jaws for numbers, and examine the chuck body for the number locations into which they fit.

The majority of lathes sold today are equipped with reversible master jaws. These jaws are held in place with socket head cap screws. When changing the jaws, be sure that the cap screws are installed in the same holes that they came out of. The long side of the jaw will typically have a longer bolt than the short side. If the bolts are not installed correctly, the jaw will not be securely fastened into place and may become dislodged while rotating. Use your best common sense when working around any piece of equipment, especially those that are bigger and heavier than you are.

One other advantage of having master jaws is that top-jaws can be

Figure 6.10 The reversible top jaws in this 3-jaw chuck are held in place with socket head cap screws. Be sure to keep the fasteners in their correct location, or damage to the receiver threads may result.

Figure 6.11 *This chuck is fitted with master jaws.*

mounted on them. Again, top jaws allow the lathe to hold an odd profile workpiece. Job shops find this especially attractive because of the increased surface and holding area the top jaws allow. Standard and reversible jaws are not very wide and do not provide the necessary support for thin or flimsy workpieces. Excessive pressure from any set of jaws can distort the workpiece to a point that the geometry of the machined parts will be distorted.

A 3-jaw chuck not found in some shops is the self-centering rotating chuck. It mates to a Morse taper adapter with a ball or roller bearing interface, and the bearing allows the chuck to rotate with the workpiece. This chuck makes an incredible difference when machining tubing, or when holding a long, heavy workpiece, as would be the case when turning an automotive crankshaft for polishing.

Three-jaw chucks are available in the following sizes: 3", 4", 5", 6", 8", 10", 12", 16", 20", 25", 32", and larger.

4-jaw independent chuck

The 4-jaw chuck is a simple and wonderful piece of engineering. With it, you can clamp a part or workpiece of almost any shape. At first glance, the complete adjustability of the jaws makes this style chuck appear complex and confusing because each of the jaws must be moved with some precision to accurately center the workpiece, and this can take a certain amount of patience and practice.

The 4-jaw chuck may be shipped with two sets of jaws: one for external clamping (the most common) and the other for internal holding. Some manufacturers machine the jaws for use in either direction (*Figure 6.12*). This is a real plus because you don't have to keep track of the extra jaws.

It is interesting to note that many machinists prefer 4-jaw chucks over 3-jaw types. The main reason is to reduce wasted time. This may seem

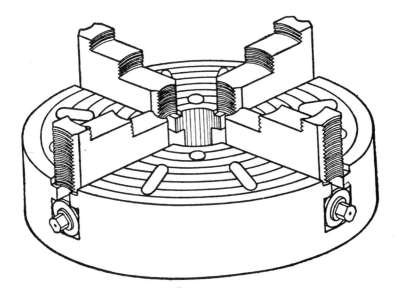

Figure 6.12 A 4-jaw chuck. Note that the jaws can be used in either direction.

contradictory to the warning about patience and practice being needed, but once you become proficient with the 4-jaw you may not want to use a 3-jaw again. Remember that in the machining business, time is money: on large machines, it takes several minutes to change from one chuck to another. Another reason for using a 4-jaw is its accuracy; 3-jaw chucks will eventually wear out and become inaccurate. This is evidenced by part runout. Using shims to get a part to run concentric in a 3-jaw chuck is also a huge waste of time. Get to know how to use the 4-jaw chuck. You may find that you will prefer it to the 3-jaw types.

6-jaw chuck

Like the 2-jaw chuck, the 6-jaw is not a part of every machine shop's inventory. This chuck is used where more *even* clamping of the workpiece is required. The six jaws support and clamp the part every 60°. Jaw widths are relatively narrow because there are so many of them.

Figure 6.13 *A 6-jaw chuck.*

These chucks are supplied with standard jaws, reversible jaws, and top jaws. Again, top jaws allow you to machine workpieces with very complicated shapes. Typical 6-jaw sizes include: 4", 5", 6", 8", 10", 12", and 16" diameters. Other sizes may be available by special order.

Rotating chucks

The use of rotating chucks is becoming more commonplace today when machining long workpieces to tight tolerances. The chuck is held in the tail stock with a Morse taper mount. A relatively new feature found on some rotating chucks is the ability to center the chuck to the workpiece to within a 1° margin. Typically, rotating chucks are found within 3-jaw configurations, but 4-jaw chucks can also be installed to facilitate the turning of off-center work. Self-centering rotating chucks are a big help when machining tubing, or when holding a long, heavy workpiece.

Chuck Keys

The chuck key is one of the most abused and overlooked tools used on the lathe. Don't get into the habit of leaving the chuck key in the chuck. This is another one of those beginner's mistakes that can cause serious damage to the machine—not to mention the operator, or someone else in the vicinity of the lathe. If the key is left in the chuck and the lathe is accidentally turned on, the key can fly off the machine, or it might grab your clothing and pull you into its path. The key could actually stay in the chuck and jam against the lathe bed. If the switch was turned on in reverse, the chuck could come undone from the spindle. So, <u>always</u> remove the key.

To prevent accidentally leaving the key in the chuck, self-ejecting chuck keys were developed. You cannot leave them in the chuck because they have a spring-loaded sleeve that ejects the key from the keyhole if pressure is not applied to hold it in place. Chuck keys are available in the following sizes for square-key types: $1/8$", $3/16$", $1/4$", $5/16$", $3/8$", $7/16$", $1/2$", $9/16$", $5/8$", $11/16$", $3/4$", and sometimes larger. Handle lengths will vary in size, but they will never be so long that you could overtighten the chuck, which can cause irreparable damage that can shut down your lathe until the necessary repair parts arrive. Even worse, for some models of chucks, repair pieces may no longer be available.

Chuck Safety

Due to chuck rotating speeds and cutting forces generated during the machining process, great care should be taken to ensure the chuck is used properly and safely. Cleaning and maintenance should be performed often for safety and to extend the life of the chuck. Keep in mind the following safety recommendations.

READ the chuck installation, operation, and maintenance manual.

DON'T start the lathe until all is clear.

REMEMBER, a collision between the chuck and other parts of the lathe will definitely do some sort of damage to both.

DON'T use the chuck on heavy work where the chuck jaws project appreciably from the chuck body. Use the correct size chuck for the job.

DON'T clamp long workpieces in the chuck without additional support. This can cause heavy damage to the lathe and work environment.

ALWAYS remove the chuck wrench before starting the machine.

DON'T tamper with the chuck. If inaccuracy is found (runout), check the spindle nose or transition adapter plate for true-running and make sure there is no dirt or foreign matter between the mounting faces.

NEVER exceed the maximum safe speed (rpm) of the chuck. Maximum rpm is generally stamped onto the face of the chuck.

NEVER operate the chuck if any parts are damaged, missing, or cracked.

INSPECT and service all chucks periodically to compensate for wear.

BE SURE top jaws are securely bolted to master jaws.

NEVER perform any unauthorized chuck modifications.

ALWAYS keep the chuck clean and well lubricated with the proper grade and quantity of grease recommended by the manufacturer.

5-C Collet Chucks

What should you do when you have to machine a pile of small parts by yourself? One suggestion would be to use a 5-C collet. The 5-C collet is akin to the chuck, but it holds smaller diameter workpieces, and you don't necessarily have to twist a chuck key to tighten it. You will need a collet adaptation chuck (like the one in *Figure 6.14*) and a collet closer (which is discussed below). The 5-C collets come in many sizes, from $1/64$" to $1\ 1/8$". Most have round holes for round stock, but some have square holes, and others have hexagonal holes for six-sided stock. 5-C collets squeeze the workpiece gently and uniformly, and provide a good deal of support.

Another method of mounting collets on any spindle is to use a scroll style collet adaptation chuck, as shown in *Figure 6.15*. It uses a chuck key to close the collet around the part. This setup may not be as fast acting to open and close as a collet closer, but it does speed up the workflow. In addition to allowing quick mounting of the workpiece, collets provide a high degree of accuracy, and are preferred when the job being performed calls for high precision.

Initially, 5-C collets were developed for speed lathes (sometimes called "*chuckers*"), but today they can be used on all lathes thanks to the availability of adaptation chucks. While other styles of collets are used for workholding on lathes, the 5-C is by far the most popular.

The majority of 5-C collets are made from case hardened mild steel. They can hold diameters ranging from a few fractions of an inch to $1\,{}^1/_8"$ (28.6 mm). The beauty of any collet is its ability to hold stock of varying

Figure 6.14 *A plain collet adaptation chuck for an A style (short taper) spindle nose. (Courtesy of Hardinge Inc.)*

Figure 6.15 *A scroll style collet adaptation chuck. (Courtesy of Hardinge Inc.)*

diametrical sizes, within limits of course. Machine stock, as received from the manufacturer, may have slight irregularities in diameter when it is produced. Typically, stock will be on the large side, not the small. Therefore, the collet must be able to handle these small discrepancies with ease. If you choose the right size collet, the jaws of the collet will spring-open slightly larger than the workpiece diameter. This allows the stock to move in, out, and through the collet. The over (plus) and under (negative) size range of a collet is typically .010" (0.25 mm).

Round hole 5-C collets are available in a wide range of sizes. Most suppliers offer them in sizes from $1/64$" to $1\ 1/8$", in increments of $1/16$". They are internally threaded and have a thread size of $1\ 3/16$"–24.

Metric round hole 5-C collets are usually found in sizes of 0.5 mm to 27.0 mm, in increments of 0.5 mm.

Figure 6.16 This chucker lathe uses a 5-C collet to hold the workpiece.

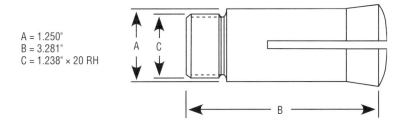

Figure 6.17 *Typical dimensions of a 5-C collet. A = overall diameter. B = overall length. C = thread size.*

Square hole 5-C collets are typically found in a range of sizes from $1/8$" to $3/4$", in increments of $1/16$". Many sizes are externally threaded to provide additional internal area for work pass-through.

Hexagonal hole 5-C collets normally come in sizes ranging from $1/8$" to $7/8$" in increments of $1/16$", and all sizes are usually externally threaded to allow work pass-through.

Collet Closers

Collet closers are used to increase production rates and reduce operator fatigue. The outside dimensional size of stock that passes through the closer is limited by the opening of the collet, but length is not a problem because special feed tubes can be placed at the end of the lathe to keep a steady supply of material flowing through the machine.

Closers may be either *manual* or *pneumatic* in operation. Manual closers are opened and closed with a handwheel or lever at the left side of the head stock, and the spindle must be stopped before loosening the collet. Pneumatic closers use an air-operated piston to pull the collets open and closed.

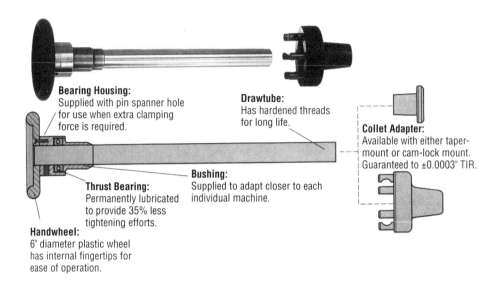

Figure 6.18 A handwheel style collet closer. (Courtesy of Royal Products.)

Collet closers are an option on most types of lathes, except the speed lathe where it's standard equipment. Collet-closer kits can be installed on most 9" to 14" swing lathes. The determining factor is the spindle bore. Sufficient spindle bore diameter must exist in order to install the device. A handwheel-operated manual collet closer is shown in *Figure 6.18*.

Cross-Slide and Compound-Rest

Cross-slide

Some lathes are constructed so that you can mount fixtures or other types of tooling to the compound-slide. This is a useful feature for some workpiece mountings. The fixture holds the workpiece and the chuck holds the cutting tool, allowing the lathe to be used as a drill press or as a horizontal boring mill.

Compound-rest

Certain production runs may require that the compound-rest be removed to facilitate the installation of a fixture to hold the workpiece for machining. This can be useful when a limited number of machines are available to produce parts.

The T-slot in the compound toolpost allows a variety of tooling to be mounted, including quick change tooling, turret toolholders, and special fixtures used to hold the workpiece. The main advantage that the compound holds over the cross-slide is its ability to place the tool or work at an angle.

Steady rest

On many lathes, a steady rest is used to support long workpieces. Without it, the parts would bend during the cutting process. The steady rest can be mounted on the bed's ways on either side of the saddle to provide the necessary support to machine a long piece of stock.

Figure 6.19 The steady rest is used to support long workpieces.

The workpiece is held by a chuck on the spindle end, and supported by the steady rest at a selected position along its length. The rest contacts the workpiece on bronze bearing points or rollers. Take special note if the workpiece has a rough surface texture. None of the support bearings (bronze points or rollers) will work well on a rough surface. An extra machining step may be required to turn a smooth area about an inch or so wide to facilitate the placement of the support bearings.

Installation is not difficult. Once the steady rest is positioned on the lathe bed, ensure that the clamp bar on the bottom of the rest is correctly positioned and tighten it with a wrench. Take care when lifting the steady rest, as it may be pretty heavy. You may need to use a crane or get help to lift it into place. Adjustment is not overly difficult. Place a level onto the length of the workpiece to ensure that it remains level during the installation process.

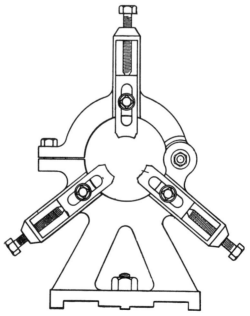

Figure 6.20 Details of a steady rest.

Follower rest

The follower rest is similar in design to the steady rest, except that the follower rest is attached to the saddle—not the bed. This allows the rest to follow or track the part. The follower rest is especially important when turning long, spindly workpieces. The natural tendency of the part to push away from the cutting tool is eliminated because the follower rest follows closely behind the cutting tool and steadies the workpiece.

Installation of the follower rest is simple—it bolts onto the right side of the saddle. Adjustment is performed with two handles or knobs that move the support bearings toward or away from the workpiece. Exercise care when positioning the support so that you do not influence the alignment of the workpiece. If care is not taken, it can be forced out of alignment with the chuck and tail stock centerline.

Mandrels

A mandrel is a shaft or spindle on which a workpiece may be fixed for rotation. It can be used for positioning hollow workpieces between centers. Expansion mandrels have a very gradual taper (.006" per foot) over their

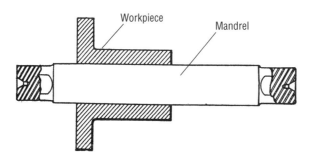

Figure 6.21 The mandrel shown in this illustration allows the entire workpiece to be machined—there is no lathe dog or chuck jaw to interfere with the cutting tool.

length, a flat on each end for securing it to the spindle, and a center drilled hole on each end for mounting on centers. When using these mandrels, you must be certain that the workpiece is squarely wedged into position on the mandrel, and it is advisable to preoil the mandrel beforehand to aid in disassembly. Plain mandrels are similar in appearance to the tapered variety, but their diameter is constant along their length. Other mandrels have a more pronounced taper—one of these can be positioned at each end of the workpiece to provide support—while others have a threaded end with a nut that can be screwed in until it contacts the workpiece.

Taper Attachment

In order to cut an accurate and consistent taper into a long workpiece, a taper attachment is required. Taper attachments are not always standard equipment on engine lathes, but they can be retrofitted without too much difficulty.

The attachment is clamped or bolted to the backside of the lathe bed, with a bolt attaching it to the cross-slide which provides the motive force to cut the taper. The cross-slide nut must be disengaged before operating the taper attachment, then the amount of taper can be set and the attachment does the rest. When the carriage moves, the attachment moves the cross-slide the amount necessary to cut the desired taper.

Taper attachments reduce setup time, and are much more accurate than using an offset tail stock to cut tapers. Also, the part can be held in a chuck while cutting the taper, which greatly benefits accuracy.

Chapter 6 - Workholding Devices 153

Figure 6.22 A taper attachment (top) and taper indicator on the attachment (bottom).

Review Questions for Chapter 6

1. A threaded spindle nose is usually reserved for _____ lathes.

    A. Heavy duty

    B. Smaller

    C. Larger

    D. None of these.

2. A cam-lock spindle nose uses a series of _____ to keep the chuck engaged to the spindle.

    A. Cam operated locks

    B. Wedge operated locks

    C. Threaded nuts

    D. Threaded screws.

3. A flat bottomed workpiece can be bolted directly to the _____.

    A. Cam lock sprags

    B. Tail stock Morse taper

    C. Face plate

    D. Wedge-lock clamps.

4. To hold a workpiece between centers, you must first
   _____.

   A. Mill flats on each end of the workpiece
   B. Center drill each end of the workpiece
   C. Clamp one end of the workpiece in a chuck
   D. Support each end with a bull nose.

5. Ball bearing centers _____.

   A. Rotate with the workpiece
   B. Carry heavier loads
   C. Reduce friction
   D. All of the above.

6. Chucks will have from _____ to _____ jaws.

   A. 1 to 2
   B. 2 to 3
   C. 2 to 4
   D. 2 to 6.

7. A universal chuck uses a _____ to move the jaws all at the same time.

   A. Scroll gear
   B. Ratchet gear
   C. Slip-ring
   D. Keyed spline.

8. Two-jaw chucks will typically use _____ jaws to clamp the workpiece.

   A. Bottom
   B. Top
   C. Intermediate
   D. Serrated.

9. A 4-jaw independent chuck is different from a universal design, because _____.

   A. Each jaw moves separately
   B. All jaws move at the same time.

10. Chuck keys should be _____.

    A. Used to open and close the chuck jaws

    B. Never left in the chuck

    C. Never abused

    D. All of the above.

11. 5-C collets are available in both _____ and _____ sizes.

    A. Inch and metric

    B. Hex and round shape

    C. Large and small

    D. None of these.

12. The compound rest is best used for making _____ cuts.

    A. Facing

    B. Drilling

    C. Tapered

    D. None of these.

13. A steady rest is used for _____.

    A. Machining short small-diameter parts

    B. Machining long large-diameter workpieces

    C. Holding the tail stock tight against the ways

    D. None of these.

14. Taper attachments are used to cut _____.

    A. Tapers

    B. The face of the workpiece

    C. Grooves

    D. Chamfers.

CHAPTER 7 Part Alignment

In this chapter we will take a look at how the workpiece is aligned on the lathe, and the necessary tools for the job. Depending on the job, alignment of the workpiece may or may not be of great importance, but someone will have to make that decision—most likely it will be you.

Types of Alignment

First, let's define what alignment is and how it pertains to the job at hand. In simple terms, alignment is the relationship between the centerline, or axis, of the stock and the machine spindle. Engineers use terms like *radial* and *axial* runout to describe the properties of alignment.

Radial (sometimes called conical) alignment refers to whether or not the workpiece is *concentric* to the centerline of the spindle. When a part is concentric, it rotates around the center of its *axis*, which, in our case, will be the rotational axis of the spindle. A part that does not rotate around its rotational axis is considered eccentric. If a part rotates around its central axis, it is concentric and void of radial runout.

Axial runout is more of a problem than radial runout. Let's use an example to describe axial runout. A piece of stock 15" long is placed into a 3-jaw chuck. Approximately 3" of the stock is held in the chuck, while the remaining 12" hangs out the end. When we check the stock nearest the chuck, we find that it is running concentric; however, when we check the unsupported end, we find that the stock is eccentric by .100". This can indicate that the stock is bent, or the chuck jaws are misaligned. In a nutshell, this is axial runout: one end runs true, but the other does not.

In any case, we want our stock, or workpiece, to run concentric to the spindle, unless we intend to machine a special offset into the piece. The tools used to align parts in the lathe can be found in any machinist's toolbox.

Alignment Tools

Alignment tools include a ruler, dial indicator, test indicator, plunger-back indicator, magnetic base, and a spirit (or "bubble") level.

Ruler

The ruler is used for approximate measurements because it's not classified as a precision tool. A ruler is handy for rough measuring of center offsets when turning tapers between centers, or when checking for lathe swing and length capacity. Machinists use quality metal rulers, not wooden rulers that can easily be damaged along their measuring scale. Here is the procedure for using a ruler to set the offset between the head stock and the tail stock.

Setting lathe center offsets with a ruler.

1. Install centers in both head and tail stocks.

2. Move the tail stock near the head stock. Leave enough gap between the two for the width of the ruler.

3. Place the ruler at a 90° angle to the center axes.

4. Move the tail stock the amount desired to produce the specified taper, e.g., $1/4$", $1/2$", $5/8$", etc., as measured on the ruler scale.

5. Lock the tail stock into place.

Dial indicators and magnetic bases

There are several versions of standard "dial indicators" and holders. All dial indicators have a graduated face that looks very much like the dial on a clock. On most indicators, the numbers on the dial begin at zero, which is usually placed at the top center of the dial, and continuously increase in value clockwise around the dial face. However, "balanced" faces like the one shown in *Figure 7.1* can be used to indicate movement either toward or away from the indicator. Although some sophisticated dial indicators have digital read outs (DRO) that provide unmatched accuracy, we are going to pay more attention to manual analog indicators—they offer high accuracy at relatively modest expense.

The "standard" dial indicator is activated by a plunger at the base of the dial. As the plunger moves inward, the linear amount that the plunger moves is recorded on the dial. Based on the length of the plunger and the mechanics of the indicator, the total amount of movement that can be measured may be very small or relatively large. Some indicators can only record very small movement, and have a range of 0 to .015". At the other extreme, indicators with long plungers are available that have a range of

0 to 12". All dial indicators should have a mounting bracket for attaching it to a base or holder, or a magnetic back for mounting it directly onto any magnetic surface.

Test indicators look very much like standard dial indicators, but they are comparative instruments that are very useful for machine setups. There are two basic varieties: plunger style and lever style (see *Figure 7.3*). The lever variety can be somewhat more difficult to use as the indicator moves in an arc rather than in a straight line, so you should become very experienced with its capabilities before using it for setups—it can be tricky for a beginner. The advantage of the lever style is that it is more adaptable to smaller, controlled working areas. On plunger style test indicators, the plunger is usually located in the back. These instruments are very useful and versatile, and usually have a measurement range of 0 to .100", in graduations of .001". Both styles of test indicators use holders for mounting to a fixed surface or magnetic base.

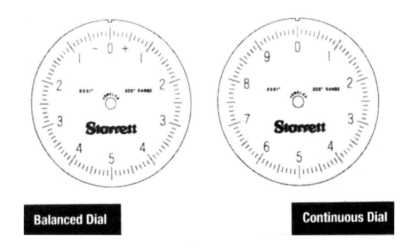

Figure 7.1 Balanced (left) and continuous (right) indicator dials. (Courtesy, L.S. Starrett Co.)

Chapter 7 - Part Alignment   **163**

A magnetic base is an essential tool for lathe setup. It can be attached to any iron-base surface, such as those on the cross-slide, the compound, and the toolholder. Some magnetic bases have a pendent arm onto which the indicator can be fastened. Others have a straight rod sticking out of the middle of the base. With a variety of clamps, you will be able to mount a dial indicator to the base and adjust it to any number of positions to help you complete your setup. It is also possible to purchase and install magnetic backs for dial indicators. They permit the indicator to be installed directly onto the bed of the lathe to track precise movement of the carriage. This feature alone makes their acquisition a "must have" addition to your toolbox. The minimum indicator travel should be no less than 1", and no more than 2". It's true that 3" and 4" indicators are available, but the plunger safety shield makes them somewhat unwieldy because of their extra length.

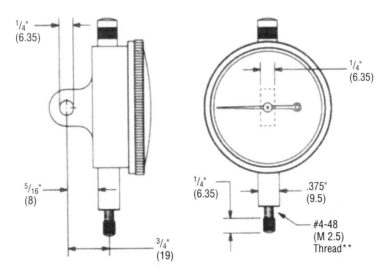

Figure 7.2 A standard manual dial indicator. (Courtesy, L.S. Starrett Co.)

164 Lathe Operation and Maintenance

Dial-type indicators are used for numerous measurement and setup operations on the lathe. As described below, they can be used for length measurement and to check a workpiece for runout.

Length measurement with a dial indicator.

1. Determine the starting point for the tool bit by placing the bit on the workpiece at the point where the machining operation will begin.

2. Place a dial indicator (with sufficient plunger travel to record the length of the cut) onto the bed ways on the left side of the carriage, and adjust the dial to read ZERO.

3. Start the machining process. Keep your eye on the indicator—it will record the exact length of carriage travel, which will be equal to the length of cut on the workpiece.

Axial runout measurement with a dial indicator.

1. Locate the indicator on the workpiece and adjust the dial until a ZERO reading is obtained.

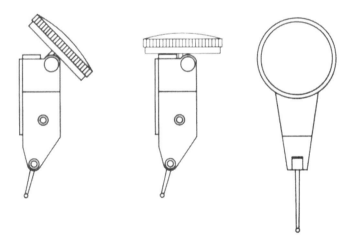

Figure 7.3 Lever style test indicator with swiveling head. Plunger style test indicators very similar in appearance to standard dial indicators. (Courtesy, L.S. Starrett Co.)

2. Rotate the chuck by hand. Note any deviations, plus/minus from the ZERO point as the part is turned. The deviation is the amount of runout.

Note: If sufficient space is available, check both ends of the part for axial runout.

Workpiece alignment with a plunger-back indicator (4-jaw chuck).

1. Clamp the workpiece into the 4-jaw lathe chuck. The workpiece should be snug, but not tight.

2. Affix the plunger-back indicator to a magnetic base. The magnetic base should be located on the machine cross-slide. Position the indicator so that the dial faces you.

3. Feed the cross-slide toward the workpiece until the indicator attains a .030" reading. This should be sufficient for most alignment operations.

4. Rotate the workpiece until the lowest indicator reading is observed. ZERO the dial at the lowest reading.

5. Rotate the workpiece to the highest indicator reading. This reading should be exactly 180° from the lowest reading. The reading on the indicator dial shows the amount of off-center, or eccentricity, the workpiece is misaligned.

NOTE: You, or your supervisor, will have to make the final decision as to the amount of misalignment that is permissible. Most machinists will allow .001" to .002" of misalignment, however, some parts must be held to ZERO. Learn to shoot for the ZERO mark.

NOTE: With square or odd shaped parts, it may be necessary to temporarily move the indicator out and away from the workpiece before it can be rotated. If the indicator must be moved, ZERO the cross-feed dial so that the indicator can be returned to the exact same location as it was before the move. Remember to remove the backlash from the cross-feed dial before you ZERO and move the feed handle.

6. Tighten the chuck on the high side until the reading is one-half the high reading, e.g., if the high reading was .028", adjust until the dial reads .014". This should bring the high side to a ZERO reading: if not, make further adjustments until an acceptable amount of eccentricity is reached.

Spirit levels

Spirit, or bubble, levels are valuable for leveling your machine, and they can also be used in some setup procedures. Precision machinist's levels are available and should be used if at all possible, as they will provide reliable operation. The best of these precision levels provide "10-second vial accuracy," which means that if the level is placed on a .0005" per foot incline, the bubble on the vial will move .100" (just under $^1/_8$ inch). Beware of carpenter levels that can be purchased at any hardware store. They will not provide the degree of accuracy required for machining.

There are two primary uses for a level on a lathe—machine setup and part setup. As would be expected with any piece of heavy equipment, a lathe should be leveled prior to use and operation, although small benchtop lathes may not require leveling because of the short parts that they will produce. However, the bigger the lathe, the more important it is to level the machine. The lathe must be adjusted end-to-end and side-to-side.

When the building that your lathe resides in was constructed, someone was in charge of making sure that the foundation was level, and, when the building slab was poured, it probably was level to a certain degree. With age, however, some buildings will become off-level—perhaps not much, but enough so that you will have to compensate for the inaccuracy when setting up the lathe.

The lathe should rest on metal or hard rubber pads. Simply making adjustments with an adjusting screw directly onto the concrete will only cause the concrete to eventually fracture underneath the adjusting screws.

When long workpieces are machined on a lathe, the need for a level will be immediate. Rather than guess if the part is parallel to the bed throughout its length, place a spirit level on the part and make any necessary adjustments. The use of a level will make short work of the whole process.

Machine setup with a spirit level.

1. Place the spirit level lengthways onto the bed ways to check for end-to-end alignment of the lathe.

2. Make corrections as needed. Be sure that each adjusting screw is carrying its share of the weight.

3. Place the spirit level across the bed to check for front-to-back alignment of the lathe.

4. Make corrections as needed. Be sure that each screw is carrying its share of the weight.

5. Recheck the lengthways alignment. Make necessary adjustments as required, until the machine is level.

6. Recheck the front-to-back alignment. Make necessary adjustments as required, until the machine is level.

Part setup with a spirit level.

1. Install a steady rest on the bed of the lathe, and position it as near as possible to the point at which the workpiece cutting zone terminates.

2. Feed the workpiece through the steady rest and clamp the other end into the chuck.

3. Place a spirit level onto the workpiece.
4. Adjust the lower steady rest support bearings until the workpiece is level with the lathe bed.
5. Adjust the top bearing until the part rotates freely and does not bind.

Review Questions for Chapter 7

1. Tools used for alignment include _____.

    A. Dial rulers

    B. Dial bore gages

    C. Test indicators

    D. None of these.

2. The two primary types of indicator faces used for part alignment are the _____.

    A. Balanced and unbalanced dial types

    B. Balanced and counterbalanced dial types

    C. Continuous and balanced dial types

    D. None of these.

3. When aligning a workpiece in a 4-jaw chuck, the part should be _____.

    A. Clamped snug, not tight, during initial setup

    B. Rotated to find the middle indicator reading

    C. Rotate the part 90° and tighten the chuck on the top side

    D. None of these.

4. What can be set up on a lathe with a spirit level?

   A. The lathe bed and the motor alignment
   B. The lathe bed and part setup
   C. The lathe bed and saddle
   D. None of these.

CHAPTER 8
Cutting Parameters and Tool Geometry

Cutting tools have a finite effective life span, which means that replacement or resharpening must be performed periodically. Tool life, while predictable to a degree, is influenced by workpiece materials and by several decisions that must be made by the machinist performing the operation. Factors that must be taken into consideration include: hardness of the material to be cut, the depth of cut, chip load, abrasiveness of the material, and the amount of time it will take to perform the machining process. We will soon examine feeds, speeds, and the other machining parameters, but first let's consider the types of cuts used on the lathe.

Types of Cuts

Basically, all cuts made on a lathe can be divided into two categories: *continuous* and *interrupted*. In a continuous cut, once the tool enters the workpiece it remains in the cut without interruption from one location to another. This is the simplest cutting that you will perform, and it is also the best for longevity of the cutting tools.

The interrupted cut is just that—a cutting action where the tool bit starts and stops in the same tool path. Let's use a couple of examples to illustrate the difference. If we mount a piece of bar stock into the lathe and cut (turning the workpiece) it from one end to the other, then we are performing a continuous cut. However, if we mount a pipe flange that has six mounting holes drilled through the flange and begin removing material from the face (facing the workpiece) of the flange, we will find that the tool will not have anything to cut when it encounters the mounting holes in the flange. It will skip over the void of the hole and begin cutting again on the opposite side of the hole. This start–stop cutting action is called an interrupted cut, and it is undesirable for both the cutting tool and the workpiece.

The biggest problem that is encountered when making an interrupted cut is that the surface finish of the workpiece will likely be marred. If the tool in the cut is not properly supported, the surface of the workpiece will become uneven, especially where the tool leaves the void area and returns to the workpiece surface. Even under ideal conditions, you may find that the tool wants to skip and hop on the workpiece surface, causing an uneven surface finish. When machining interrupted cuts, reduce the recommended feed and speed rates, and keep the tool bit short and well supported to reduce chatter and to avoid visually unacceptable surfaces.

Photomicrographs reveal that cutting tools do, in fact, jump and skip on the workpiece surface. When the tool makes its initial contact with the surface, it bounces, and because the part is revolving, when the tool starts its downward arc from the first bounce, it will likely bounce again. Simply put, reduce the amount of time spent making interrupted cuts by machining as much of the part as possible before adding holes and/or grooves to the workpiece.

Speeds, Feeds, and Depth of Cut

The term "cutting speed and feed" encompasses a lot of information. You have probably already heard references to "speeds and feeds," and the meaning is not as obvious as it may first appear.

Cutting speed

Cutting speed, used in this context, does *not* refer to how fast the spindle is revolving, and it is *not* expressed in revolutions per minute (rpm). Instead, cutting speed is measured in surface feet per minute (sfm or sfpm), or surface meters per minute (smm or smpm), and it is a measurement of the distance, in feet or meters, that a point on the circumference of the workpiece travels (rotationally) in one minute. Therefore, a cutting speed of 120 surface feet per minute (120 sfm) means that if you made a mark at some point on the workpiece, that mark would travel (rotate) a distance of 120 feet in one minute.

Another way to look at sfm is to imagine a round piece of stock with a 120-foot-long piece of string attached to it. If we spin the stock fast enough to roll up the entire length of the string in one minute, the surface speed would be 120 sfm. When we remove stock from a workpiece, we are essentially removing that string—the string, of course, being the length of the chip(s) we remove. But how do you determine how many rpm the spindle should be turning to make the mark travel 120 feet in one minute? Here is how the rpm can be calculated, given the circumference or diameter of the workpiece.

First, the diameter of the workpiece is measured, and then its circumference is calculated from the measurement. For example, if the diameter is 3.25 inches, the circumference (which is the distance around the workpiece), in feet, can be found with this formula:

Circumference = ($\pi \div 12$) × diameter in inches

Circumference = (3.1416 ÷ 12) × 3.25 = .85 feet.

Now that the circumference is known, the required rpm can be found with this formula.

rpm = SFM ÷ circumference

rpm = 120 ÷ .85 = 141 rpm.

From this example, you can see that a 3.25 inch diameter workpiece (this is sometimes referred to as cutter diameter, rather than workpiece diameter, but the terms are interchangeable) would have to run at 141 rpm to achieve a surface speed of 120 sfm.

Before moving on, the following formula is also helpful. It allows you to determine what the sfm will be when you know the rpm and workpiece circumference.

sfm = circumference × rpm.

Using the speeds and dimension from the above example, you can determine the sfm that would result from cutting a 3.25 inch workpiece with a circumference of .85 feet at 141 rpm.

sfm = .85 × 141 = 119.85, or rounded up to 120 sfm.

These formulas can be used for all turning applications on the lathe. Make sure that you learn them now before you move on in this chapter.

Cutting feed

Cutting feed is measured in inches per minute (ipm) or millimeters per minute (mm/m), and it is the distance the tool moves along the workpiece from one point to another (a *linear* movement) in the time of one minute; sfm, remember, is a *rotary* measurement. Measurement of ipm is pretty simple. Put a nonpermanent mark (with a marker pen) on the way just in front of the saddle. Start the machine table feed and allow the saddle to travel for exactly one minute. Make a mark at the stopping point, then reverse the carriage and measure the distance between the first and second mark. That distance, in inches, is ipm.

You will quite often see feeds expressed as inches per revolution (ipr) or millimeters per revolution (mm/r), especially on machining charts that provide recommended feeds for cutting. This is a measurement of the distance the tool moves along the workpiece for every complete revolution of the workpiece (or spindle). Inches per minute can easily be converted to inches per revolution with the following formula.

ipr = ipm ÷ rpm.

There are two classes of material removal that are used in machining most workpieces: roughing and finishing. Roughing is the rapid removal of stock without regard to surface finishing. Finishing is the last cut that is made on the workpiece, using a relatively shallow depth and light feed to obtain precise dimensions and smoother surfaces. Generally, roughing cuts on an engine lathe are made at feed rates ranging from 0.010 to 0.025", and finishing cuts are made with the feed set in the range of 0.003 to 0.010". The actual feed will depend on many variables including the material being cut and the cutting tool material used to make the cut.

Changing the feed can have a dramatic impact on productivity. For example, doubling the feed rate also doubles the chip thickness (which makes the chip more difficult to curl and bend), but at a cost: the power required to drive the spindle increases, as does the cutting temperature. Tool life is also decreased. Clearly, deviating from the recommended feed rate will impact several cutting parameters, and it should not be done without considering the consequences.

Depth of cut

Depth of cut, which is often abbreviated as d.o.c. (or simply doc), refers to the amount that the cutting tool penetrates the workpiece. It is normally measured in hundredths or thousandths of an inch, or in tenths or hundredths of a millimeter. On a lathe, the depth of cut is equal to one-half of the total amount of material removed from the workpiece in one revolution. This was explained earlier, when we compared machining on a lathe to skinning an apple—equal amounts of material are removed from both sides of the workpiece.

Ideally, when turning a workpiece, it is advisable to try to complete the operation in two cuts: one roughing cut that will remove the bulk of the material, and one finishing cut that will remove considerably less material, but will leave a smooth surface finish. Of the three cutting parameters (speed, feed, and depth of cut), depth of cut is the easiest to alter. Doubling the depth of cut will double productivity without increasing temperature, cutting force per square inch of material removed, or the strength of the chip. However, it will also double the amount of horsepower consumed and increase the rigidity requirements of the lathe, so it is very important to confirm that adequate power and machine rigidity are available before increasing d.o.c.

Summary

All materials have recommended cutting speeds, as can be found in machinists' reference materials, or as published by the material manufacturer. The references presented in *Table 8.1* will suit the majority of general applications. Note that these are conservative speed recommendations for both HSS and carbide tooling, based on a feed of .012" ipr and a .125" d.o.c. A range of speeds is provided for each materials group—if you are unsure of hardness of the material being machined, begin with the lower speed and slowly increase sfm until the best speed for your machine is found. Through experimentation, you will discover just how much a tool can be pushed.

Using inappropriate speeds and feeds will ultimately shorten the life span of any tool. Running a tool too slowly will not only increase the machining time, but will also increase heat buildup in the tool. Running it too fast will not allow the tool sufficient time to remove material in a satisfactory manner and will result in a rough surface finish. There is a delicate balance between the speeds and feeds and the cost of doing business. Increasing speed will increase tooling costs and require the machinist to stop and change tooling more often. Decreasing speeds will tie up the machine for longer periods of time and increase labor costs.

Finally, remember that all of the information provided above on speeds, feeds, and depth of cut can be thrown out the window if your cutting tools

Table 8.1: Recommended Cutting Speeds.

Material	Cutting Speed (sfm)	
	HSS tool	Carbide tool
Machine Steel	50 – 150	150 – 450
Tool Steel	50 – 70	150 – 210
Cast Iron	30 – 120	90 – 360
Bronze	90 – 100	270 – 300
Brass	150 – 200	450 – 600
Aluminum	500 – 800	1500 – 2000

Based on a feed rate of 0.012" per revolution and 0.125" depth of cut.

are not sharp! Every suggested speeds and feeds chart you will encounter assumes that you are using sharp tools. Never begin a cutting operation with a dull cutting tool—the results will be either disappointing or disastrous.

Cutter Geometry

Now that we have looked at the basic cutting parameters, let's look at tool geometry and cutting forces to provide a better understanding of what influences the success of a turning operation. The effectiveness of the cutting tool is affected by the angle of the tool in relation to the workpiece. How the tool performs while making the cut is directly influenced by its geometrical relationship with the workpiece. Using the best cutting angle can greatly improve any lathe operation.

Top rake angle (also known as the back rake angle or angle of inclination) is shown in *Figure 8.2*. It is the angle formed between the angle of inclination of the tool bit and the line perpendicular to the workpiece

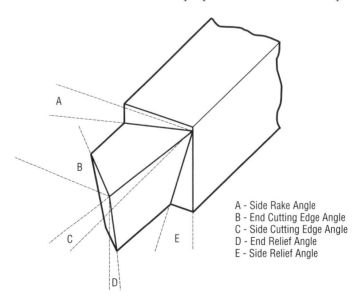

A - Side Rake Angle
B - End Cutting Edge Angle
C - Side Cutting Edge Angle
D - End Relief Angle
E - Side Relief Angle

Figure 8.1 *Single point tool terminology.*

when viewing the tool *from the side*, front to back. The top rake is positive when the angle slopes downward from the cutting point and into the shank, and negative when it slopes upward from the cutting point and above the shank. It is neutral when the line formed from the top of the insert is parallel with the top of the shank. It should be noted that inserts are also designed to be either positive or negative, as shown in *Figure 8.3*.

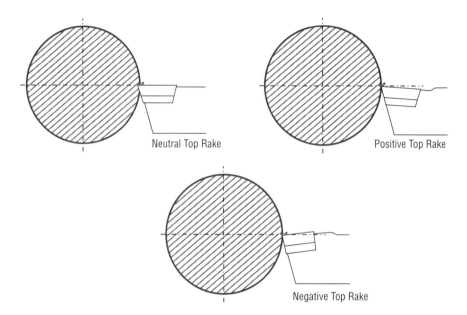

Figure 8.2 Top rake angles, viewing tool from the side.

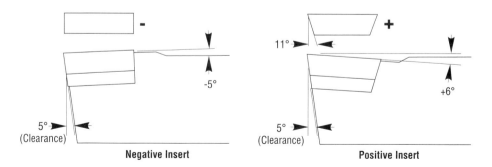

Figure 8.3 Negative and positive insert application.

Side rake angle is formed between the face of the insert and the line perpendicular to the workpiece when viewing the tool from *the end*. It is positive if it slopes away from the cutting edge, and negative if it slopes upward. It is neutral when perpendicular to the cutting edge. The thickness of the tool behind the cutting edge is dependent on the side rake angle. A small angle allows for a thick tool that can provide maximum strength, but will require higher cutting forces *(Figure 8.4)*.

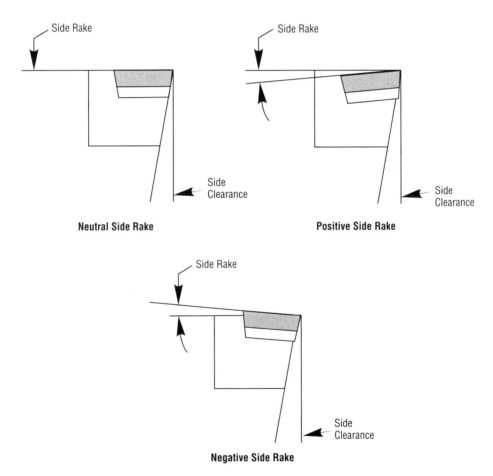

Figure 8.4 Side rake angles, viewing tool from the end.

Increased rake angles produce thinner chips, reducing cutting force requirements. However, once a maximum recommended angle is exceeded, the cutting edge is weakened and heat transfer is diminished. *Table 8.2* provides suggested top rake angles for selected workpiece materials.

Table 8.2: Top Rake and Clearance Angles for Selected Workpiece Materials.

Material	Recommended Rake		Recommended Clearance
	Carbide	HSS	
Aluminum	20° positive	25° positive	10°
Titanium, Inconel	10° positive	25° positive	6°
Low Carbon Steel, Ductile Brass	5° positive	25° positive	5°
Malleable Cast Iron <160 BHN	5° negative	18° positive	5°
High Carbon Steel, Cast Irons >160 BHN	0–5° negative	12° positive	4°
Steel Castings	10° negative	12° positive	6°

Side cutting edge angle or *lead angle* is formed between the tool's side cutting edge and the side of the tool shank. This angle leads the tool into the workpiece. Enlarging this angle produces wider chips. Side and end cutting edge angles are shown in *Figure 8.5*.

End cutting edge angle is the angle formed between the insert cutting edge on the end of the tool and a line perpendicular to the backside of the tool shank. It determines the clearance between the cutting tool and the finished surface of the workpiece.

End relief angle is formed below the end cutting edge by the end face of the tool bit and a line perpendicular to the base of the toolholder. The end clearance angle may be greater as it is formed by the end face of the toolholder and a line perpendicular to the base of the toolholder. Tip overhang will make the clearance angle exceed the relief angle. (See *Figure 8.1*.)

Side relief angle is formed below the side cutting edge by the side face of the tool bit and a line perpendicular to the base of the toolholder. The side clearance angle may be greater as it is formed by the side face of the toolholder and a line perpendicular to the base of the toolholder. Tip overhang will make the clearance angle exceed the relief angle.

Relief and clearance angles on the end, as well as the side, of the cutting tool are required to permit the tool to enter the cut. Without clearance, it would be impossible for chip formation to occur. Without adequate relief, the cutter will rub and produce heat. However, too large an angle results in a weak cutting edge. In general, the softer the workpiece material, the greater the top rake of the turning tool. Softer workpiece materials require high positive shearing angles, while harder, tougher materials are best cut with more neutral or negative geometries. Cutting tools also need end and side

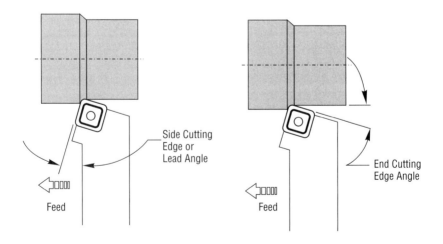

Figure 8.5 Side and end cutting edge angles. (Courtesy of Kennametal Inc.)

clearance angles in order to enter the cut. Tools without adequate clearance will push away from the workpiece and there will not be sufficient clearance to allow chip formation to take place.

Tangential, radial, and axial cutting forces are all at work in a turning operation. The tangential force has the greatest effect on the power consumption of a turning operation, while the axial (feed) force exerts pressure through the part in a longitudinal direction. The radial (depth of cut) force tends to push the workpiece and toolbar apart. When these three components are added together, a resultant cutting force is established. The ratio of these forces is approximately 4:2:1 for a zero degree lead angle tool. This means that if the tangential force equals 1000 lb (454 kg), the axial (feed) force will be 500 lb (227 kg), and the radial (depth of cut) force will be 250 lb (113.5 kg). The magnitude of the radial and axial forces is altered

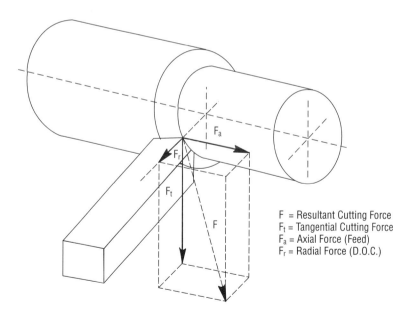

Figure 8.6 Cutting forces in turning operations. (Courtesy of Kennametal Inc.)

as the toolholder lead angle is changed. In a turning operation, the greater the lead angle, the larger the radial cutting force and the smaller the axial force *(Figure 8.6)*.

Side cutting edge angle (also known as lead angle or attack angle)

The primary consideration when selecting the side cutting edge angle (lead angle) of a turning tool is often the geometry of the workpiece or the material condition (see *Figure 8.5*). For example, when cutting through scale, interruptions, or a hardened surface, a lead angle tool will allow a reasonable rate of productivity without subjecting the cutting tool edge to severe shock. This benefit must be balanced against the possibility of

Table 8.3: Lead Angle/Chip Thickness Relationship.

If the Lead Angle is	And Feed Per Revolution (fpr) is	Then, Actual Chip Thickness * is
0°	A	A
10°	A	0.9848 × A
15°	A	0.9659 × A
20°	A	0.9397 × A
30°	A	0.8660 × A
45°	A	0.7071 × A
Examples Using fpr Values		
0°	0.010"	0.010"
10°	0.010"	0.009848"
15°	0.010"	0.009659"
20°	0.010"	0.009397"
30°	0.010"	0.00866"
45°	0.010"	0.007071"

* Chip thickness (ct) is the cosine of the lead angle multiplied by the feed per revolution (ipr). ct = cosine (of lead angle) × ipr.

increased part deflection or vibration due to the higher radial force created by the lead angle. The radial force increases as the lead angle increases. The majority of turning operations are most efficiently performed within a lead angle range of 10° to 30°.

Chip thickness is equal to feed when using zero degree lead angle cutting tools. However, when lead angle cutting tools are applied, chip thinning occurs. *Table 8.3* shows how lead angle changes affect chip thickness.

Example 1. Where lead angle is 20°, and ipr is 0.010", find the chip thickness. (The cosine of 20° is .9397.)

0.9397 × 0.010 = 0.009397"

Example 2. Calculate the difference in feed rate, in inches per minute, for a given chip thickness, between a 0° lead angle cutter and a 45° lead angle cutter. The feed rate is 0.003" ipr, and the spindle speed is 1057 rpm. (Note that the feed rate can be *increased* by 41% by changing only the lead angle.)

For 0° lead angle: in./min = ipr × rpm.

0.003 × 1057 = 3.171

For 45° lead angle: in./min = (ipr ÷ 0.707) × 1057

(0.003 ÷ 0.707) × 1057 = 4.485

Chip thickness decreases with increases in the lead angle. However, for a given chip thickness, a 45° lead angle will increase both chip width and feed rate by 41% over that achieved with a zero degree lead angle tool, all other parameters being equal.

Zero degree lead angle turning tools produce chips that are equal in width to the cutting depth of the turning operation. When a lead angle cutting tool is introduced, the effective cutting depth and corresponding chip width will exceed the actual cutting depth on the workpiece. *Table 8.4* calculates how different lead angles affect feed rate, chip thickness, and chip width in relation to increases in the lead angle.

Table 8.4: Selected Machining Parameters Influenced by Lead Angle.

Lead Angle	Feed (ipr)	Actual Chip Thickness	Chip Width
0°	100%	100%	100%
10°	101.5%	98%	101.5%
15°	103.5%	97%	103.5%
20°	106.4%	94%	106.4%
30°	115.5%	87%	115.5%
45°	141.4%	71%	141.4%

Actual chip width for any depth of cut and lead angle can be calculated from the percentages in *Table 8.4*.

Example: Where lead angle is 30°, and d.o.c. is 0.20", find the chip width (c_w).

c_w = d.o.c. × c_w % expressed as a decimal

c_w = 0.20 × 1.155 = 0.231".

Machinability

The machinability of a workpiece material can be defined as the comparative ease with which a material removal operation can be performed upon it relative to the same operation on a benchmark material. A material's rating can help in the selection and adjustment of speeds and feeds for unfamiliar materials. For example, SAE-AISI resulfurized carbon steel 1212 has a machinability rating of 100%, while SAE-AISI 1030 has a machinability rating of 70%. This means that if 3 in.^{3} (49.16 cm^3) of SAE-AISI 1112 can be removed per minute with a specific tooling setup, that same setup should remove 70% of that amount, or 2.1 in.^{3} (34.41 cm^3), of 1030 carbon steel. *Table 8.5* provides comparative machinability ratings for most materials you will encounter.

Table 8.5: Machinability Rating Guide. *(See text for explanation.)*

SAE-AISI Designation	Rating %	SAE-AISI Designation	Rating %	SAE-AISI Designation	Rating %	SAE-AISI Designation	Rating %
Carbon Steels							
1005	45	1019	78	1034	70	1046	57	1069A	49
1006	50	1020	72	1035	70	1049	54	1070A	49
1008	55	1021	78	1037	70	1050	54	1074A	45
1010	55	1022	78	1038	64	1050A	70	1075A	45
1011	53	1023	76	1039	64	1053	54	1078A	45
1012	55	1025	72	1040	64	1054	54	1080A	42
1013	53	1026	78	1042	64	1055A	51	1084A	45
1015	72	1029	70	1043	57	1059A	51	1085A	42
1016	78	1030	70	1044	57	1060A	51	1086A	42
1017	72	1031	70	1045	57	1064A	49	1090A	42
1018	78	1033	70	1045A	72	1065A	49	1095A	42

(Continued)

Table 8.5: Machinability Rating Guide. *(Continued)* *(See text for explanation.)*

SAE-AISI Designation	Rating %	SAE-AISI Designation	Rating %	SAE-AISI Designation	Rating %	SAE-AISI Designation	Rating %	SAE-AISI Designation	Rating %
1106	79	1117	91	\multicolumn{6}{l}{}			
1108	80	1118	91						
1109	81	1119	100						
1110	81	1120	81						
1115	81	1125	81						
1116	94	1126	81						

SAE-AISI Designation	Rating %	SAE-AISI Designation	Rating %	SAE-AISI Designation	Rating %	SAE-AISI Designation	Rating %	SAE-AISI Designation	Rating %
				Resulfurized Carbon Steel					
				1132	76	1141A	81	1151	70
				1137	72	1144	76	1151A	81
				1138	76	1144A	85	1211	94
				1139	76	1145	66	1212	100
				1140	72	1145A	78	1213	136
				1141	70	1146	70	1215	136
				High Manganese Carbon Steel					
1524	66	1536	64	1548	55	1552	49	1566A	49
1527	66	1541	57	1551	54	1561A	49	1572A	49
				Carbon Steel (Leaded)					
10L18	92	10L45	66	10L45A	84	10L50	60	10L50A	78
				Free Machining Steel (Leaded)					
11L17	104	11L41	79	11L44	87	12L13	170	12L15	170
11L37	84	11L41A	94	11L44A	98	12L14	170	—	—

(Continued)

Table 8.5: Machinability Rating Guide. *(Continued) (See text for explanation.)*

SAE-AISI Designation	Rating %	SAE-AISI Designation	Rating %	SAE-AISI Designation	Rating %	SAE-AISI Designation	Rating %	SAE-AISI Designation	Rating %
41L40A	77	41L50A	70	\multicolumn{2}{c}{Alloy Steel (Leaded)}					
				86L20A	77	—	—	—	—
1320A	57	1330A	60	\multicolumn{2}{c}{Manganese Alloy Steel}					
				1335A	60	1340A	57	1345A	57
				\multicolumn{2}{c}{Molybdenum Steel}					
4012	78	4037A	72	4142A	66	4340A	57	4640	66
4017	78	4047A	66	4145A	64	4419	78	4718	60
4023	78	4118	78	4147A	64	4615	66	4720	60
4024	78	4130A	72	4150A	60	4620	66	4815A	51
4027	66	4137A	70	4161A	60	4621	66	4817A	49
4028	72	4140A	66	4320A	60	4626	60	4820	49
				\multicolumn{2}{c}{Nickel Chromium Molybdenum Steel}					
8615	70	8625	64	8637A	70	8650A	60	8720	66
8617	66	8627	64	8640A	66	8653A	56	8822	64
8620	66	8630A	72	8645A	64	8655A	57	9255A	54
8622	66	8635A	70	8647A	60	8660A	54	9260A	51

(Continued)

Table 8.5: Machinability Rating Guide. *(Continued)* *(See text for explanation.)*

SAE-AISI Designation	Rating %	SAE-AISI Designation	Rating %	SAE-AISI Designation	Rating %	SAE-AISI Designation	Rating %
\multicolumn{8}{c}{Chromium Steel}							
5015	78	5130	57	5140A	70	5150A	64	5160A	60
5060A	60	5132A	72	5145A	66	5152A	64	E51100A	40
5120	76	5135A	72	5147A	66	5155A	60	E52100A	40
\multicolumn{8}{c}{Chromium Vanadium Steel}							
6102	57	6118	66	6145	66	6150A	60	6152A	60
\multicolumn{8}{c}{Alloy Steel – Boron}							
50B44A	70	50B50A	70	51B60A	60	94B17	66	94B30A	72
50B46A	70	50B60A	64	81B45A	66	—	—	—	—
\multicolumn{8}{c}{Stainless Steel}							
301	55	308	27	317	35	403	55	418	40
302	50	309	28	321	36	405	60	420	45
303	65	310	30	330	30	410	55	430F	65
304	40	314	32	347	40	416	90	440	50

(Continued)

Table 8.5: Machinability Rating Guide. *(Continued)* *(See text for explanation.)*

SAE-AISI Designation	Rating %	SAE-AISI Designation	Rating %	SAE-AISI Designation	Rating %	SAE-AISI Designation	Rating %	SAE-AISI Designation	Rating %
A2, A3, A4	16	D5, D7	11	H24, H25	15	O1, O2, O7	16	S1, S2, S5	20
A6, A8, A9	16	H10, H11	20	H26, H42	15	06	38	T1	14
A7	11	H13, H14	20	M2	14	P2, P3, P4	25	T4	11
A10	27	H19	20	M3	11	P5, P6	25	Ta5	8
D2, D3, D4	11	H21, H22	15	M15	8	P20, P21	22	W (All)	30

Tool Steel

Other Materials (Non steel)

Material	Type	Rating %	Material	Type	Rating %
Cast and Malleable Iron	Soft Cast	81	Cast Steel	BHN 120	85
	Medium Cast	64		BHN 220	50
	Hard Cast	47		BHN 245	44
	Malleable Iron	80-106	Bronze	Al Bronze	60
Brass	Yellow	80		Mn Bronze	60
	Red	60		Ph Bronze	40
	Leaded	280		Ph, Leaded	140
	Red, Leaded	180		Si Bronze	60

(Continued)

Table 8.5: Machinability Rating Guide. *(Continued) (See text for explanation.)*

Material	Type	Other Materials (Non steel)			
		Rating %	Material	Type	Rating %
Aluminum	2S	300-1500	Magnesium	Dow H	500-2000
	11S-T3	500-2000		Dow J	500-2000
	17S-T	300-1500		Dow R	1150
Nickel and Nickel Alloy	BHN 135	26	Copper and Copper Alloy	Cast	70
	Monel, Cast	45		Rolled	60
	Monel, Rolled	55		Everdur	60
	Monel "K"	35		Everdur, Leaded	120
	Inconel (B)	35		Gun Metal, Cast	60

"A" indicates annealed. Carbon Steel 1212 (100% rating) is the comparison material for ratings. *Courtesy of DoAll Company and Texaco.*

Review Questions for Chapter 8

1. Which of the following contribute to tool wear?

    A. Part diameter

    B. Part length

    C. Abrasiveness of the material

    D. None of these.

2. Cutting speed is measured in _____.

    A. Subcutaneous feet per minute

    B. Surface feed per hour

    C. Surface feet per second

    D. Surface feet per minute.

3. Cutting feed is measured in _____.

    A. Subcutaneous inches per minute

    B. Surface feed per hour

    C. Inches per minute

    D. Surface inches per minute.

4. Running a cutting tool too slow can cause _____.

   A. Less heat and more accurately machined parts
   B. More accurately machined parts and decreased machining time
   C. Decreased power consumption and less heat
   D. Increased power consumption and increased machining time.

5. Using an HSS tool bit will require cutting speeds to be _____ _____ by a factor of four as compared to using a carbide tool.

   A. Increased
   B. Decreased
   C. Remain the same
   D. None of these.

6. Name the five angles used to describe cutting tool angle terminology.

   A. _____
   B. _____
   C. _____
   D. _____
   E. _____

7. The ratio of tangential, radial, and axial cutting forces are _____:_____:_____.

   A. 1:2:4
   B. 2:4:6
   C. 4:2:1
   D. 6:4:2.

8. Diagram the differences between negative and positive rake.

9. The comparative ease with which a material is removed during a machining operation is called _____.

   A. Friability
   B. Speeds and feeds
   C. Malleability
   D. Machinability.

CHAPTER 9
Cutting Fluids and Tool Cooling

Cutting tools must be kept within a well-defined temperature range in order to maintain a sharp cutting edge. Tool cooling plays a major role in the overall performance and service life of your cutting tools. Not only does the coolant remove heat from the cutter and workpiece, it also lubricates and washes away troublesome chips. Proper lubrication is essential when cutting certain materials, especially aluminum, which can form burrs (sometimes called *smearing*) that stick onto the cutting tool. If ignored, smearing will not only render a very poor surface finish, it may also allow the size of the cut to increase beyond allowable specifications.

A wide variety of coolant types are used in machining, but petroleum-based and synthetic-based fluids are the most widely accepted for lathe use. Each one has a specific purpose, and it can make or break a job if the wrong one is used. If you encounter an unusual application and are not sure which fluid to use, contact your local coolant dealer or supplier for a specific recommendation.

Care must be exercised when handling coolants on or off the machine because most coolants are classified as hazardous waste materials once they have served their usefulness. Upon completion of a job using water-based solutions, be sure to remove all of the coolant from the surfaces of the machine to prevent rust and corrosion.

Coolant Application

There are four general types of coolant application that can be used on the engine lathe: drip, flood cooling, mist cooling, and high pressure. Drip is one of the most common, and least complicated, methods used to deliver the coolant to the workpiece. Brushes used for this job are typically inexpensive and need not be of high quality. It is important to learn how to apply the coolant properly. Apply the coolant onto the surface of the workpiece and allow it to flow as needed. At first, you may find yourself applying too much

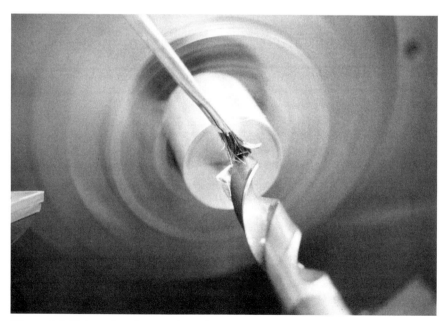

Figure 9.1 Using a brush to apply coolant.

or too little coolant, and it may take some time to predict the direction the coolant will flow on the rotating workpiece, but don't worry, your technique will improve with experience. Also, remember that placing the brush against the sharp cutter bit will result in the bristles being cut off, and using a brush with very short bristles in this manner may cause the handle to get stuck between the cutter and the workpiece, jerking it from your hand and possibly cutting your fingers or hands on the brush's metal handle in the process.

Flood cooling uses a low-pressure pump to convey coolant to the workpiece. If the proper flow of fluid is maintained, excess coolant will flow off the workpiece and into a drip tray. (However, as we discussed in Chapter 2, you can create quite a mess by applying too much fluid to a revolving workpiece—so be careful to get it right.) A drain hole is located at one end of the drip tray to allow the coolant to return to the storage and coolant tank

Figure 9.2 *Flood cooling can be used with most every turning operations.*

where the pump is located. Coolant can then cool down and be filtered to remove contaminants before it is recirculated. Flood coolant is normally restricted to low speed spindle applications as high speeds encourage the coolant to be splashed away from the machine rather than gathered in the dip tray for reuse.

Mist applicators are very popular for lathe operations, primarily because they do not create as much of a mess as flooding. Also, the workpiece is partially cooled by the air stream carrying the coolant, which means that much less liquid coolant can be used to perform comparable cooling and lubrication functions. Since mist coolant systems use one or more small wand-like nozzles to aim and direct the coolant flow, a constant supply of air pressure is required, and multiple spray nozzles may be necessary when high chip loads are expected. The most obvious drawback with mist cooling is that the coolant is released into the atmosphere where it

Figure 9.3 *Mist coolers are used with higher spindle speeds.*

can be inhaled by anyone in the vicinity. Therefore, in most instances there should be some type of vapor vacuum system to remove stray mist. You are advised to check with your coolant supplier for specific recommendations about safety precautions. Some coolants are better suited to mist cooling than others.

High-pressure coolant applicators can dramatically increase production and tool life. These applicators compress the coolant to pressures over 200 PSI (13,6 BAR) and direct the spray at the workpiece through small diameter coolant passages within the toolholder or tool. This highly pressurized coolant literally blows the chips away from the workpiece, resulting in increases in both tool life and workpiece surface finish. The downside is the overspray. To utilize this system, the machine must be equipped with fully enclosing side and top shields to trap the overspray, plus a system for

Figure 9.4 *A common spray bottle is a convenient and inexpensive alternative for applying coolant.*

collecting the large amounts of fluid. High-pressure coolant applicators work very well on computer numerical controlled machining centers with automatic tool changers, and where operator interface is very limited.

One simple way to administer coolant for light machining on an engine lathe is with a simple spray bottle, especially one with an adjustable nozzle that will allow it to apply fluids as either spray or mist. It is not only inexpensive, but also allows the coolant to be applied when and where it's needed.

Coolants

In a perfect world, we would not have to use coolants. Their use increases operational costs, and many of them must be treated as hazardous wastes upon disposal. However, for machining, their attributes far outweigh their negatives, and by learning as much as possible about what coolants are available, along with environmental guidelines for their disposal, you will be sure to get the most for your money.

As we mentioned, coolants are classified as either petroleum-based or synthetic-based. The petroleum-based types are further classified as mineral- or animal-based. The mineral types are refined from petroleum-based stock and receive special fortifiers during the refining process to enhance their interaction with the tooling and workpiece. The majority of mineral-based coolants are applied in pure (or 100%) form; so they do not have to be mixed or reduced in strength. These coolants are best applied with a brush or as a flood coolant.

Animal fat-based pastes and liquids were among the first coolants used by machinists. In fact, animal lard is still used for certain materials, especially aluminum. Application is normally performed with a brush.

Synthetic coolants are pretty much the standard for today's machine shop. They provide excellent cooling and lubrication, and in most cases can be applied in less than pure form. Mixing ratios and strengths must follow the manufacturer's recommended specifications for desirable results. The application of the coolant aerates the fluid, which adds oxygen to the solute and keeps the mixture fresh. Some mixed coolants, if not aerated regularly, will become rancid and spoil (the telltale sign is the bad odor). When selecting a coolant, be sure to check the biostability of the mix and determine how long the mixture can be stored without going bad. The objective is to mix small quantities as needed, not large quantities for storage. Water-based coolants can be applied by all methods previously discussed. Again, discuss your needs with the chemical supplier for best results.

If you plan to perform tapping operations with the lathe, you will want to use a tapping fluid specially formulated for that purpose. These oils penetrate into the workpiece and tool to make thread cutting much easier than with standard coolants, and they improve the surface finish of the threads. Thread tapping fluids are supplied in a liquid or paste-like lube, and both can be applied with a brush, or straight from the can.

Chemical and Coolant Handling

Always follow the manufacturer's handling recommendations for any type coolant. Avoid direct physical contact with your skin and eyes. Use suitable gloves, as necessary, when handling any chemical, <u>and</u> don't forget those safety glasses!

You should always know where the Material Safety Data Sheets (MSDS) for the fluids are kept, or ask for them when you purchase coolant. If the unthinkable does happen, take the MSDS with you to the emergency room so that the medical staff will know what they are dealing with. A material safety data sheet file should be set up for all chemicals used in the shop.

For those coolants that must be mixed, mix only that amount that is required for the job. Storage of mixed coolants is a bad idea. Once mixed, coolants start an immediate reaction with air. Left unattended, the mixture will spoil. Spoiled coolants are very easy to identify—they stink very badly. Spoiled coolants should not be reused for cutting, and you should limit your personal contact with them.

Most coolants are classified as hazardous waste after they have served their useful life. To that end, make sure that they are handled and transported within local, state, and/or federal guidelines.

Figure 9.5 Tramp oil skimmers and roll-type filters are used to remove oil and debris from water-based coolants.

Tramp Oil Skimmers

During the operation of any machine, a certain amount of oil and lubricant is used to keep the machine running properly. These lubricants are applied during routine maintenance. This is all well and good for the machine, but not for water-based coolants, because when petrochemical coolants mix with water-based types, their effectiveness as a coolant diminishes. There is also the problem of disposal. The cost of hauling mixed waste is substantially higher than for a single component type. Tramp oil is the petrochemical oil that floats on water—oil has a specific gravity in the .70 range, compared to that of water at 1.00.

Tramp oil skimmers use this physics-related phenomenon to their advantage. The floating oil attaches itself to a rotating disc, which is raised out of the water, and is then scraped off into a receptacle where it is held for proper disposal.

Magnetic Swarf Removal

Along with any machining process come chips—some large and some small. The majority of coolant pumps can deal with small chips during the recirculating process, and, depending upon the type of pump, the small chips are ground even smaller. Unfortunately, when left in the system these small chips cause wear in several areas. The pump impellers wear out prematurely, and the cutting tools are dulled because they have to cut and recut the same stock over and over again.

Built somewhat like the tramp oil skimmers, magnetic swarf separators use strong permanent magnets to attract small magnetic particles. The magnets are flanked by a pair of discs fitted with scrapers that remove the magnetic particles clinging to the discs, and then move them on to a waiting receptacle for eventual safe disposal.

Figure 9.6 *This magnetic particle remover extracts ferrous materials from the coolant. Swarf removal lengthens cutting tool life.*

What about nonmagnetic particles? If you were paying attention during your high school science class, you probably remember that nonmagnetic particles DO NOT collect themselves onto magnets. Therefore, you can't use a magnetic separator to remove them from the coolant.

However, you can use a three-tier coolant tank to separate the swarf from the coolant. By having three tanks, each feeding the other, what comes out of the last tank is, by and large, fairly clean. However, each tank must be cleaned from time to time to remove the remnants for proper disposal. Be sure to wear proper safety gear when handling any type of coolant(s).

Chapter 9 - Cutting Fluids and Tool Cooling **207**

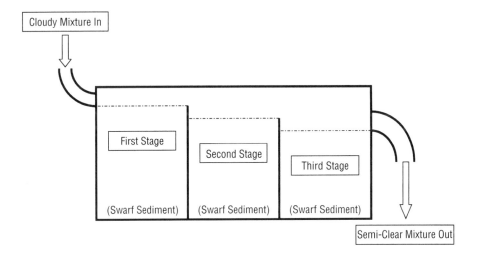

Figure 9.7 *A three-tier coolant tank can be used to remove nonmagnetic swarf very effectively.*

Review Questions for Chapter 9

1. Cutting fluids are used to _____.

    A. Increase overall cost of machining

    B. Reduce temperature and provide lubricity

    C. Reduce temperature and increase machining time

    D. None of these.

2. Coolants are applied using _____.

    A. A brush, or flood and vacuum application

    B. A brush, a spray mist, or with a flood application

    C. Only flood application

    D. Only a brush.

3. _____ and _____ are the most acceptable cutting fluids used on a lathe.

    A. Viton and copolymers

    B. FEA and NURBS

    C. animal fat and vegetable oils

    D. None of the above.

4. Spoiled coolants can be distinguished by their _____ smell.

   A. Bad
   B. Sweet
   C. Vinegar
   D. None of these.

5. Tapping lubricants _____ the surface of the workpiece to make cutting easier.

   A. Flood
   B. Penetrate
   C. Coat
   D. None of these.

6. Animal fats are still a viable cutting fluid for _____.

   A. Aluminum
   B. Tool steel
   C. Resulfurized steel
   D. Carbon based stainless steels.

7. What safety precautions should be used when handling chemicals and coolants?

   A. Wear leather cloves, grounding strap, and canvas apron

   B. Wear rubber gloves, grounding strap, and gold jewelry

   C. Wear rubber gloves and apron, and wear safety glasses or goggles

   D. None of these.

8. What does MSDS stand for?

   A. Material Safety Data Specifications

   B. Material Safety Data Sheets

   C. Manufacturers Safety Data Specifications

   D. Machinists Safety Data Sheets.

CHAPTER 10
Basic Cutting Procedures

Backlash

So far we have discussed a number of methods and tools that can be used to mount and/or hold a workpiece for machining. Now, we want to turn our attention to the basic machining operations: facing, turning, parting, drilling, boring, reaming, grooving, and threading. You will need to pay attention to the wide array of tools, and how they are mounted and positioned during each machining operation, even though many of the cutting operations will be very similar in size and scope. Before we look at the actual cutting operations, however, we need to briefly turn our attention to the lathe's mechanical controls. Almost every lathe suffers to some degree from *backlash*—one of the enemies of making perfect parts—and you will need to understand how to compensate for backlash before you can repeatedly produce acceptable parts on the lathe.

Lathes use a nut and leadscrew arrangement to move the cross- and compound-slides back and forth and in and out. There is by necessity a certain amount of play (or slack) between the nut and leadscrew. This is necessary to allow for the screw to pass through the nut and not bind as it is turned. Over time, as the machine ages, the clearances between moving parts become larger, and more play is introduced. This increased clearance means that the nut and screw are no longer in the same proximity to each other as when new, and the resulting increase in clearance is felt in the controls and is known as backlash. Adjustments can be made to lessen the effects of backlash, but it can never be totally eliminated.

And the effects will be obvious. Turning the cross-slide feed handle on a new precision lathe will result in almost instantaneous movement of the slide, but on an older lathe you may find that the handle must be turned $1/4$ to $1/2$ of a turn <u>before</u> the slide starts to move. Consider this example: let's assume that you turn the feed wheel $1/4$ turn and the cross-slide moves forward .100". Now, turn the dial in reverse $1/4$ turn, to its original position. You might logically assume that the slide has also reversed .100" to its original position. Chances are, you would be wrong. In reality the cross-slide may have only moved a total of .090", or perhaps not at all, due to the slack between the nut and screw. This is caused by backlash.

To eliminate backlash, always reverse the feed dial/wheel past the initial start point (typically ZERO) and return it to the original setting. This will remove the backlash.

The majority of work performed on the lathe will be performed near the head stock. This alone will cause the leadscrew to wear more in this location than it does at the middle and tail stock end. Any adjustments to the nut that are made to remove the backlash at the head stock end will result in the controls becoming stiff when machining near the tail stock. So, in many cases, it may be better to learn to compensate for backlash than to try to make adjustments to repair the problem.

One effective way to reduce backlash is to remove the standard leadscrew and nut arrangement and install *ball-screws*. Ball-screws use a very different approach to backlash adjustment/compensation; they use a series of balls (under constant pressure) that circulate around the screw. This constant pressure is what eliminates the backlash. Ball-screws are not usually installed as standard equipment, but are a very worthwhile option for high-precision machining and reduced machine maintenance.

On all of the operations that will be discussed below, remember that the final dimensions of the part will not be accurate if you do not eliminate backlash before making adjustments to your lathe. As discussed earlier, a dial indicator or testing gage can be a great help in measuring machining movements on the lathe. If you are not sure that you have mastered the technique of taking up play and eliminating backlash, a dial indicator can provide the reassurance you need.

Facing

Of all the cutting procedures performed on the lathe, *facing* is one of the most popular. Simply put, facing is the machining of a flat surface on a workpiece. Common applications include auto brake rotors and flywheels. In addition, facing is used to remove rough surfaces and burrs from a workpiece.

Facing, plus other operations commonly performed on the lathe, are shown in *Figure 8.1*.

We have, thus far, discussed the tools and procedures used to mount work on the lathe. Using the methods that we have discussed to this point, you should be able to mount and align the stock to the spindle. Facing is performed with the cross-slide, using hand or power feed to move the tool across the face of the workpiece.

Facing a part must be performed with the same deliberate thought used for all of the procedures outlined in this book. You must select the

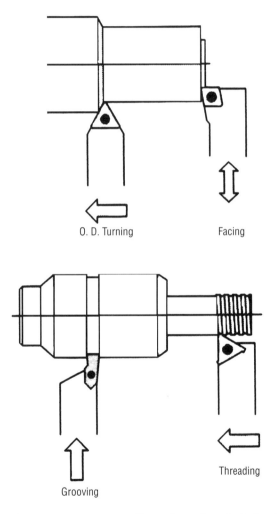

Figure 10.1 Machining operations typically performed on the lathe. (Courtesy of Kennametal Inc.)

best workholding device, toolholder, and cutting tool for the operation. For most facing, a round-nose or radius-end tool bit will be chosen because it provides the necessary amount of stock removal with the least evidence of tool path (no peaks and valleys) on the surface of the workpiece.

One difficulty involved in facing a part is caused by variations in workpiece diameter. When you start at one side and work toward a smaller or larger diameter, logic should tell you that in order to maintain a constant cutting speed (in sfm), the spindle speed should progressively *increase* when working from the outside to the center of the workpiece, and *decrease* when starting from the center and working out to the outside edge. In a perfect world, we would all have machines that could sense these requirements and change speed accordingly. Unfortunately, the majority of lathes do not have such a provision and we must work with what we have.

As a rule of thumb, use the outside diameter of the workpiece when calculating the necessary spindle speed. Unless the workpiece is very large in diameter, this will be sufficient. On larger stock you may be tempted to change speeds midway through the facing procedure, but resist the temptation. When you stop the spindle and feed, a small gouge or groove will be left in the surface of the part. All by itself, it may not be a critical factor in the operation of the part, but the surface will have an obvious blemish. Check with your supervisor or instructor before contemplating such a maneuver.

One further note: it may be possible to increase the spindle speed by as much as 25% and still leave the requisite surface finish. However, expect tool life to diminish substantially. On hardened material, stick with recommended cutting parameters.

Facing procedures

Let's take a look at some of the facing operations you may have to execute. You will find that placement of the tool bit is very important if the workpiece must be faced to the very center of the piece. Whether you start from the center or outside, you may want to ask yourself a few questions. Does it make sense to start at the middle and work outward? If I start from the outside, will the next cutting operation be within easy reach? Can I use power-feed? Answers to these questions will help determine the ultimate outcome, and they will be a factor in how long it takes to cut the workpiece.

Parts with drilled centers will pose less of a problem because the entire surface does not require cutting. A center drilled hole eliminates the requirement of the tool bit to be positioned at the exact center to the workpiece during facing operations.

Large diameter parts with a thin cross-section will be one of your more difficult challenges. Because they are relatively fragile, these parts cannot be placed within the safe confines of the chuck jaws, and you will have to use extreme caution during setup. If placed too far back into the jaws, the part cannot be machined all the way to the outer edge because the jaws will limit the travel of the tool bit.

To solve this problem, initial setup will require three small blocks of metal, all of the same thickness. (**Note:** It's very important that these three blocks are exactly the same thickness—if there is any discrepancy, the workpiece will not be parallel to the chuck face.) First, place the workpiece into the chuck and clamp it loosely into place. Then place the three blocks behind the workpiece 120° apart so that one side rests against the chuck face and the other side rests against the workpiece. Press the part against the blocks with one hand and tighten the chuck with the other. Next, remove the metal blocks. The workpiece is now *prealigned*. Remember, enough of the workpiece must be projected from the end of the jaws to allow the facing operation to proceed without interruption.

Now that the workpiece is prealigned, use your dial- or plunger-back indicator to finish the alignment process. Alignment should be within .001"–.002".

Facing becomes really tricky when both sides (or ends, depending on the configuration of the part) of the workpiece must be true (parallel) to one another. The first side is mounted and faced as already outlined. Then we must assure ourselves that the face of the chuck is absolutely flat and that it has no runout. To check this, use an indicator. Remove the jaws from the chuck. Mount the indicator onto the cross-slide or compound-rest. Check the chuck face—it should be free of runout.

Now that we are confident that the chuck will run true we can finish up our setup. Once again, use the three blocks that were used to prealign the workpiece. Place the workpiece into the chuck jaws and tighten lightly. Insert the blocks behind the workpiece. Press firmly against the part to ensure that it is flat against the chuck. Now, tighten the workpiece into place. Remove the metal blocks.

As a final check of your work, install a test indicator onto the cross-slide. Rotate the workpiece so that the jaws will facilitate the entry of the indicator onto the backside of the workpiece. Adjust the indicator point so that it will reach the backside of the workpiece. Using the cross-slide feed wheel, move and adjust the indicator on the backside of the workpiece until a ZERO reading is achieved. Rotate the workpiece to determine if it continues reading ZERO throughout a full rotation.

Caution: Don't crash the chuck jaws into the preset indicator.

If the part does not continuously read ZERO, you will need to tap the edges, preferably with a soft metal headed hammer, until it does.

If the workpiece has a hole in the middle, your setup procedure will be less time consuming as the part might possibly be held from the inside diameter. If this is so, then the indicator can be mounted to read the variation without interruption from the chuck jaws. This would be the case when aligning a flange for machining.

Turning

Turning is the term we use to describe the removal of stock along the length of the workpiece. We normally think of turning as the operation performed when the carriage moves from right to left, or vice versa. There are two types of turning that can be performed on a lathe—*straight* and *tapered*. Each type is performed on the exterior of the workpiece.

Straight turning

For straight turning, the best setup process is dependent upon the size and shape of the workpiece. For round stock, a 3-jaw chuck is preferred. Use a 4-jaw chuck on those pieces that have irregular shapes. 5-C collets and drawbars can be used when turning long pieces of bar stock. Once the workpiece is clamped and aligned, a cutting tool is selected to suit the particular shape that will be cut. Tool selection is an integral part of the setup process. Knowing if you need a right-hand or left-hand tool bit with a chisel- or round-nose point will make a huge difference in production speed.

We have already discussed the differences between a 1:1 and 2:1

feed dial. You will need to know which system your lathe uses before you begin.

If the cutting distance is small, simply hand-feeding the tool will suffice. When longer lengths must be cut, use the power-feed. The procedures used to machine the workpiece will most likely depend on the desired shape of the finished part. Sometimes, you may have to experiment to determine the order and number of steps to complete the part. Later, we will discuss how to write process and setup sheets. These sheets can speed up the whole setup and turning process. Once you are familiar with the various functions and operations of the lathe, you will get to a point where you will just automatically know what operation to complete next, but it does take time to get to this point.

Turning a part to size.

Understanding how to accurately reduce the diameter of a workpiece is an essential part of becoming a machinist. If you can't "hit" (a broad term used here) size, then your part will wind up in the scrap heap. Let's take a look at a typical sizing problem and determine how it is resolved.

In this problem, the machinist is working with a piece of SAE-1018 steel that is .625" in diameter. The part is to be machined to a finish diameter of .400" on one end. The first step the machinist must take is to determine the spindle speed required for the operation. From a chart of cutting speeds in a machinist's handbook, it can be determined that the cutting speed for SAE-1018 is 90 sfm. Using the formula we learned in Chapter 8, the diameter is converted to circumference, measured in feet.

$(3.1416 \div 12) \times .625 = .163625$ ft.

Given the circumference and the cutting speed, the required rpm can now be found with another formula we have already learned.

$90 \div .163625 = 500$ rpm.

So 500 rpm will be the required spindle speed.

Once the workpiece is mounted and the tool bit is aligned, the cross-slide feed handle needs to be adjusted to a ZERO point, and the first reduction cut can be made. The operation is initiated by turning the machine On and advancing the tool bit until it just touches the spinning workpiece. This will leave a very shallow mark that will be visible on the peripheral edge of the part. At this stage, the movable cross-slide adjustment bezel should be adjusted to ZERO.

Next, the workpiece is cut approximately .200" in length to remove all surface debris, and then the workpiece diameter is measured with a micrometer (not a caliper). For this example, let's say that the part is now .612" in diameter. If the final finished diameter (.400") is subtracted from .612", it can be determined that an additional .212" must be removed from the diameter to get the part to size. Remember, this reduction is to be .212" from the total diameter, which will mean removing .106" from each side. If the lathe has a 1:1 ratio, the total advance feed required on the cross-slide will be .212" to get the part to size; but if it has a 2:1 ratio, the cross-slide advance feed will be .106" to obtain the correct finish diameter.

Because the material being turned is a fairly hard steel, the total amount cannot be removed in a single pass. If it is, the resulting surface finish will be unsatisfactory. Therefore, it will be necessary to take several rough cuts of .050", and a final finish cut of .002". Take a look at the chart below—it demonstrates how to get from .612" down to .400". The dial movements provided on the chart are for a 2:1 lathe.

Example, using a 2:1 ratio lathe.

Cut	Amount to be Removed	Turn Dial	Measured Size
1st	.050"	.025"	.562"
2nd	.050"	.025"	.512"
3rd	.050"	.025"	.462"
4th	.050"	.025"	.412"
5th	.010"	.005"	.402"
6th	.002"	.001"	.400"

As can be seen, it will take a total of 6 cuts to get to the final size. The 5th cut is only .010" deep so that only .002" stock will have to be removed on the final finish cut.

Offset turning

Offset turning is a straight turning operation whereby the workpiece is offset from the spindle centerline and machined. A good example of offset turning would be an engine crankshaft. The rod journal(s) centerline is machined offset from the main bearing centerline. The engine's stroke is twice the offset amount, e.g., offset the crankshaft 1.500" to achieve a 3" stroke, etc. An example of a crankshaft having one rod journal and two main journals set up for turning can be seen in *Figure 10.2*. The upper drawing in the figure shows the main journals supported between centers, and the bottom illustrates how the rod journal can be turned if the main journals are offset mounted.

Short workpieces can be completely held and offset with a 4-jaw chuck. Long parts will have to be supported at both ends and will require some sort of special production fixture. Depending upon the workpiece, you may be able to center drill each end for the specified amount of offset and place it between the centers.

222 Lathe Operation and Maintenance

The offset turning procedure described below is for a short workpiece held with a 4-jaw chuck.

Offset turning with a 4-jaw chuck.

1. Center the workpiece in the 4-jaw chuck.

2. Rotate the chuck until the jaws are located at the 12:00, 3:00, 6:00, and 9:00 o'clock positions.

3. Mount a 1" travel dial indicator onto the cross-slide with the indicator tip contacting the backside of the workpiece. ZERO the gage.

4. Move the workpiece the specified amount using the 3:00 and 9:00 o'clock jaws. For this demonstration example, let's say .100" offset. The indicator dial should move .100" as the workpiece is moved toward the rear of the machine. Tighten all four jaws.

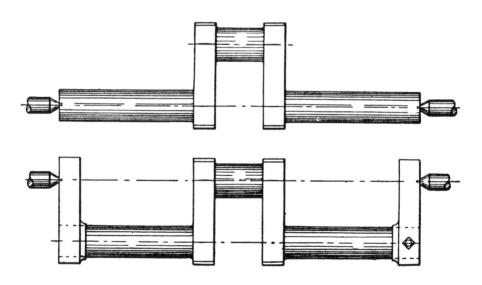

Figure 10.2 Mounting methods for turning crankshaft journals.

5. Rotate the workpiece very slowly to ensure that it is exactly .100" off-center (axis).

6. Machine the workpiece to the specified diameter. Exercise caution when machining parts off-center. Slow the machine rpm down 25–30%, and take light cuts until the journal is completely round.

Making interrupted cuts on any workpiece will typically cause more stress on the tools and holders than any other types of machining that you will perform. The offset workpiece is a worst-case scenario when it comes to tool positioning and machining practices. Be patient, and don't get in a hurry when machining offset parts.

Taper turning

Taper turning is performed either with a taper attachment or by moving the tail stock off center. However, before discussing tapers, it is important to understand the terminology used to describe a taper. *Figure 10.3* shows a finished part that is twelve inches long and has a taper of one inch per foot. The dotted line indicates the centerline of the part. Note that, even though the taper is one inch per foot, the part is two inches larger on one end than on the other. This is because taper is calculated from the centerline—the taper on each side of the centerline is one inch per foot.

Tapers can also be described by the angle created by the taper. In the illustration, angle is the *included* angle, or the angle created across the full width of the finished part. The other angle, $1/2$, is known as the angle with the centerline.

When taper turning a part, you must be certain of the specifications. As you can see, if the taper angle is specified as the included angle, but the machinist interprets it as the angle from centerline, the part will be useless.

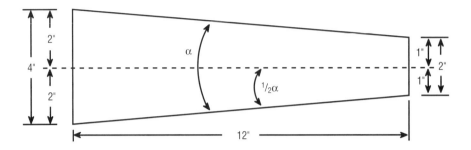

Figure 10.3 A tapered part with dimensions.

Turning a taper is a skill that every lathe operator must acquire. Understanding the dynamics of taper turning is not difficult. Tapers can be cut with a form tool, by offsetting the tail stock, and with a taper attachment.

READ the print specification carefully to determine if the angle is (or is not) an included angle.

Cutting a taper with a taper attachment.

1. Mount the workpiece into the chuck and align.
2. Adjust the carriage so that it is slightly ahead of the workpiece, and then adjust the position of the taper attachment so that the carriage will have sufficient room to move from one end of the workpiece to the other.
3. Release the cross-slide feed nut so that the cross-slide will follow the path set by the taper attachment when the carriage is moved from side to side.
4. Adjust taper attachment, in degrees or taper per foot.
5. Move the tool bit close to the workpiece and prepare to start cutting.

6. Adjust the spindle speed for the type of material that you will be cutting.
7. Turn the machine on and cut the required taper.
8. Check work to ensure that the angle was machined correctly.

Calculating tail stock offset for turning tapers.

When cutting a taper by offsetting the tail stock, the following formula can be used to provide the amount of offset necessary.

Tail stock offset in inches = $1/2 \times$ ([taper per foot $\div$ 12] $\times$ length of work).

If you do not know the taper foot, but you do have the desired diameters for each end of the finished part, subtract the small diameter from the large diameter, and divide the remainder by 2. For example, if the large diameter is 4 inches, and the small end is 2.5 inches, the tail stock would be offset $3/4$" to turn the part.

Checking the work with a taper gage.

1. Clean gage with a soft lint-free rag before use.
2. With a piece of chalk, draw a line on the workpiece parallel to the length of the taper.
3. Insert the workpiece into the gage and twist/rotate slowly one-half turn.
4. Remove workpiece from gage. The chalk mark should be completely removed or smeared on that portion of the workpiece that the gage came into contact with. If the chalk mark is still visible on any part of the workpiece, then the angle is incorrect and must be adjusted further to make necessary corrections.

Parting

Once you have machined the surfaces of the workpiece to specifications, it may be necessary to cut it off from the rest of the stock. This is done with a *parting* or *cutting-off* procedure. Parting is very important; you can ruin the whole job if the part is not cleanly removed.

Setup of the cutting tool should be performed with great care to ensure that the tool is placed exactly 90° to the longitudinal axis of the workpiece—89° or 91° won't work; you will cut either a concave or convex shape onto the bottom surface, which will probably ruin the part.

The cutting tool may require a positive or neutral rake to suit the cutting characteristics of the workpiece material. Special toolholders are designed for cut-off tools and should be used for this purpose. The actual cutting procedure is not difficult, but it does take some practice.

The spindle rpm should be set low, i.e., very low, when compared to normal turning speeds. Feed should be done by hand. Power-feed can be used, but you should have considerable experience before attempting this maneuver.

Exercise care when the part is nearly cut off. Letting it drop onto the bed or carriage could damage the surface of the part. Do not use your hand to retrieve the falling part.

By far, the most common mistake made when parting involves the feed rate. The whole process is rather slow and tedious to perform. Don't be tempted to *push* or *crowd* the tool into the workpiece. Pushing the tool with excessive force can cause the workpiece and/or cutting tool to bind. When this happens, and it most certainly will, the part may stop turning and the tool bit may shatter. The resulting force of the shattering pieces will cause surface damage to the part. Be cautious, and be careful! Start slow and increase feed and speed as the tool gets closer to the center of the workpiece. Big diameters require slow speeds and feeds, while small diameters can be parted at higher speeds and feeds.

Center Drilling

On the lathe, before a hole is drilled into a workpiece, it must be center drilled. Holes not center drilled will almost certainly be drilled off-center or eccentric to the part's rotational axis. Most holes can be center drilled with a standard length center drill. For those holes that are located in places where a standard center drill cannot reach, extra-length drills are available. Always try to use the shortest length center drill possible. This will reduce drilling errors.

The standard (or normal) angle found on center drills is 60°. This fits the live or dead center perfectly. However, it's possible to purchase center drills with 82° and 90° ends. Be sure that the center drill you use on the lathe is 60°, otherwise severe damage to the lathe center will surely result.

The 90° center drills are often used for chamfering purposes. The 82° center drill can be used to machine the recess necessary for countersunk socket head screws.

There are two other styles of center drills: the *bell type* and the *radius type*. The bell type is used to machine a center hole into a workpiece that requires a protected center. Protected center holes eliminate the risk of deformation of the center hole seat. The radius type center drill allows a faster feed rate during machining. The unique radius shape virtually eliminates center drill breakage because of the continuous radius. The radius type center drill should also be chosen for machining tapered workpieces between centers.

Grooving

Grooving is a simple machining process whereby a cutting tool is plunged into a workpiece to create a groove or channel. There are essentially three types of grooves: square-cut, round-cut, and V-cut.

One of the most common reasons to cut a groove into a workpiece is for the installation of O-ring seals. O-rings are used to seal pneumatic or hydraulic pressurized joints between two parts.

Square-cut grooves are the most typical that you will cut. Tool position is very important. Having the cutting tool at anything other than a right angle to the workpiece can cause the groove to be misaligned.

Round-cut grooves are not used for O-rings because they do not allow the proper support for the seal. In reality, the O-ring seals compress against the sides and bottom of the square-cut groove and against the connecting part. Round-cut grooves do not allow the ring to deform and take the shape necessary to form the correct seal.

V-cut grooves are used for various reasons and can take two forms: the two-sided and three-sided forms. The two-sided types require a sharp pointed tool bit ground at a suitable angle. The three-sided types essentially use the same type of tool bit, but the groove has a *floor* to form a trapezoid shape into the surface of the workpiece. The shape is produced by moving the grooving tool side to side, on or in the surface of the workpiece.

External grooves

Setup of external grooving tools is not difficult. Obtain the desired cutting tool, install into a toolholder, align to the surface of the workpiece, and cut the groove to depth using the cross-slide.

Placement of the grooving tool will depend on the tool used and the material of the workpiece. Some materials will require a positive, neutral, or negative rake tool position. Once you decide which type to use, install it into a sturdy toolholder. Try to eliminate overextension of the cutting tool—by having the tool sticking out too far, you will create a natural tuning fork. At certain spindle rpm, the tool will start to vibrate and cause small ripples in the surface of the workpiece. Consider your spindle speeds very carefully, but the rule is to keep it slow. Don't try to cut a groove at full speed; reduce speed by 50% to 70%. Once you determine the capability of your machine and tools, it will be possible to sneak up on the speed issue. When everything is right, you will be able to cut grooves at or near full speed. Full speed, in this case, is the speed at which you would perform turning.

Depth of the groove is regulated by how much you move the cross-slide feed handle. On 1:1 lathes, you will need to divide the depth by two because in most cases the dimensions for groove depth are measured from the surface of the workpiece. The tool bit must be placed against the outside surface of the workpiece, at which point the feed dial must be set to ZERO.

Measurement of the groove can be made with vernier calipers, depth micrometers, and gages manufactured for this purpose.

External grooving procedures.

1. Mount the workpiece in the lathe chuck.
2. Install and align the O-ring or V-ring cutting tool. Be sure that the tool bit is square (90°) to the workpiece. You don't want to cut a diagonal groove into the workpiece.
3. ZERO the tool against the workpiece.
4. Set the spindle speed required for the workpiece diameter, which is typically very slow.

5. Ensure that there is an adequate coolant supply to keep the tool bit cool and well lubricated.
6. Cut the groove into the workpiece.
7. Measure the O-ring groove depth, or the V-groove angle and depth to confirm accuracy.

Internal grooves

The setup used to cut internal grooves is substantially different from the one used to cut external types. The cutting tool must be mounted onto a bar small enough to allow entry into a predrilled hole in the workpiece.

Like the external groove, the internal cutting procedure for external grooves is very similar. Just remember to back the tool out of the cut and move it toward the center of the workpiece before you remove it, or you will cut the interior surface of the part.

Measurement of the internal groove is performed with tools designed for this purpose. The use of snap gages and the like will not work because they will need to be collapsed before they can be removed.

Threading

One of the primary reasons for the development of the lathe was to cut screw threads with some degree of accuracy and repeatability. The lathe allows us to cut all sorts of threads, including external, internal, and tapered forms.

There are two basic thread (customary and metric) forms for threaded fasteners: Unified and ISO. Unified thread specifications were developed in the U.S. and they are measured in customary U.S. inch units—they are specified by the diameter of the fastener and by how many threads are cut into a length of one inch. ISO (International Organization for Standardization) specifications refer to the distance, in millimeters, between each thread peak. Both types share the same basic form, which is a 60° thread angle.

Threads allow us to *fasten* two or more components together. Whether we are connecting two pieces of pipe to extend its length, or using a bolt to clamp two pieces of material together, a threaded fastener is best suited for this task and to aid in the eventual disassembly process.

Let's take a look at some of the different types of thread you may be required to cut. Standard threads have a 60° thread form that can be cut with a simple 60° threading/grooving tool. By far, this will be the most common type of thread cutting performed on a lathe. Cutting a 60° thread is certainly not without its problems, and it can be damaged quite readily if not properly handled.

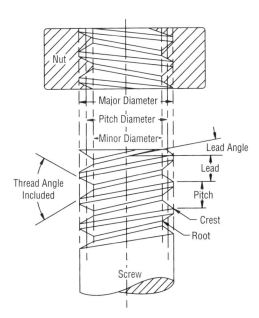

Figure 10.4 Basic screw thread nomenclature. *(Courtesy of Kennametal Inc.)*

External threads

Cutting external threads is a primary lesson in your lathe operation education. The machine setup procedure is not too difficult, but the hand–eye coordination may take a little bit of practice.

> **Some Simple Rules About the Threading Dial.**
> **Even Threads**—if the same as leadscrew or any multiple, that is, 8, 16, 32, etc., engage at will without reference to thread dial. For even threads other than above, engage on any line of the dial.
> **Odd Threads**—cut by engaging any of the numbered lines.
> **Half-Threads**—such as 4-$\frac{1}{2}$, engage on 1 and 3, or 2 and 4, but not both.
> **Quarter-Threads**—such as 5-$\frac{3}{4}$, use any mark and return to the same mark for each cut.

Procedure for cutting external threads with a lathe.

1. Set the compound-rest to 29° pointing towards the spindle.

2. Install the workpiece into the chuck.

3. Install a 60° thread cutting tool bit into the toolholder.

4. Align the threading tool to the workpiece with a setup gage.

5. Set the spindle speed to approximately 30-40 rpm.

6. Adjust the compound-slide feed dial to ZERO. Be sure that the long part of the feed handle is facing the 12 o'clock position. Remove all backlash in the counterclockwise rotation. Readjust the dial to the ZERO position.

7. Start the lathe and adjust the cross-slide handle until the cutter just touches the workpiece. Stop the spindle. Adjust the dial to the ZERO position.

8. Move the carriage to the right side of the workpiece.

9. Adjust the cross-slide feed dial .010".

Chapter 10 - Basic Cutting Procedures   **233**

10. Start the lathe. Apply a liberal amount of cutting oil to the surface of the workpiece.

11. Place your left hand on the threading engagement lever, and your right hand on the compound-slide feed handle.

12. Engage the threading engagement lever once the dial aligns with any of the numbers—not slightly before, not slightly after—right on the number!

13. Once the cutter has reached the end of the threaded area, <u>quickly</u> retract the tool bit from the workpiece by moving the compound-slide feed dial approximately $1/2$ to $3/4$ of a turn. Then, disengage the threading feed lever. This 1–2 operation will take a while to develop, but it must be learned. The hand–eye coordination that is required to cut threads is one of the most important skills that you will have to master.

> *THE threading gage should be adjusted to engage the leadscrew at this time if it has not already been done.*

> *THE positioning of your hands is critical at this point. Your left hand should be on the threading feed engagement lever, your right hand should be on the compound feed handle. You will need to learn this routine— no matter which hand you favor.*

14. Move the carriage to the right-side of the workpiece. *Caution:* be sure that the tool bit does not collide with the workpiece during this move.

15. Reset the compound feed dial back to the ZERO starting point. Notice that you don't reset the cross-slide feed handle. Resetting of the cross-slide handle would move the cutting tool out-of-phase with the workpiece, and you will lose the ability to measure how deep the cutter has moved into the workpiece.

16. Apply another liberal amount of cutting oil to the surface of the

workpiece. This must be done before and during the cutting process to ensure smooth cutting.

17. Advance the cross-slide another .010" and repeat the cutting procedure.

18. Repeat this cutting process until the threads are starting to form peaks. Advance the feed .002" to .005" per cut at this point.

19. Repeat the cutting process until the threads are cut to the specified final size.

> *THE numbers and lines on the threading dial are there to help you engage the lever and cut at the right place and right time—every time. For an even number of threads you can use any of the lines or numbers to start with. For odd-number threads, use only numbers or lines—not both.* **Note:** *Exception to the rule— if you must cut double-threads, then you will have to use both lines and numbers, otherwise you will cut off the threads you just made.*

Measurement of external threads

There are several ways to measure threads, but this section will concentrate on the two most popular methods: Go/No-Go thread gages, and thread measurement using wires.

A Go/No-Go thread gage is a precision tool that conforms to thread form specifications. The tool should be handled and stored with care to prevent it from being damaged. Threading gages may be either adjustable or nonadjustable. The adjustable types must be preset or calibrated prior to use. This will ensure consistent size checking ability. The tool is threaded onto the part as if you were installing a nut. You must *feel* the fit when using this gaging process—either too tight or too loose will result in the part being rejected.

The "over wires" measurement of threads is somewhat more involved and slightly more complicated. This gaging process is recognized the world over as being reliable and accurate, and it involves placing two wires on one side of the threaded workpiece, and one on the opposite side as shown in *Figure 10.5*. The wires provide surfaces (the tops of the wires) that can be measured with a micrometer.

For the measurement to be accurate, the wire should contact the thread at the *pitch diameter*; this is called the *best* wire size. Wires come in a variety of sizes, and you will have to select the best size to fit the threads being measured by using the following equation to identify a constant for the angle, and then dividing the constant by the pitch (number of threads per inch).

Constant for the angle = secant of $1/2$ angle of the thread × 0.5.

As can be determined from a table of trigonometric functions, the secant of 30° (which is one half of 60°, the thread angle) is 1.1547, and 0.5 × 1.1547 = 0.57735.

Therefore, for all 60° threads, this means that the best wire size is 0.57735 ÷ pitch (the number of threads per inch).

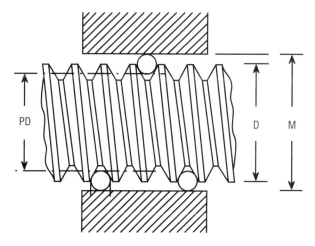

Figure 10.5 The three thread measurement system. M= Measurement over the wires. D = Outside diameter of the thread. PD − Pitch diameter of the thread.

The best wire size for measuring a $^{7}/_{16} \times 24$ Unified thread would be found by dividing the constant by the pitch, or $0.57735 \div 24 = 0.02405625$. Best wire sizes for Unified threads are given in *Table 10.1*, and best wire sizes and constants for Metric threads are shown in *Table 10.2*.

Metric constants are derived from a different formula based on pitch and the wire diameter rather than the number of threads per inch. Wires used in thread measurement should be hardened and lapped steel wires with an accuracy of 0.0002 inch.

Table 10.1: Three-Wire Measurement for Unified Threads.

Pitch	Best Wire Size	Pitch	Best Wire Size
4	0.144338	20	0.028868
4.5	0.128300	22	0.026243
5	0.115470	24	0.024056
5.5	0.104973	26	0.022205
6	0.096225	27	0.021383
7	0.082478	28	0.020619
7.5	0.076980	30	0.019245
8	0.072169	32	0.018042
9	0.064150	36	0.016037
10	0.057735	40	0.014434
11	0.052538	44	0.013121
11.5	0.050204	48	0.012028
12	0.048112	50	0.011547
13	0.044411	56	0.010310
14	0.041239	64	0.090211
16	0.036084	72	0.008018
18	0.032075	80	0.007217

All dimensions in inches.

After three wires of equal diameter have been selected, they are positioned in the thread grooves as shown in *Figure 10.3*. The anvil and spindle of an ordinary micrometer are then placed against the three wires and a measurement is taken. To determine what the reading of

Table 10.2: Three-Wire Thread Measurement for Metric Threads.

Pitch		Best Wire Size		Constant	
Millimeters	Inches	Millimeters	Inches	Millimeters	Inches
0.35	0.01378	0.2021	0.00796	0.3031	0.01193
0.40	0.01576	0.2309	0.00909	0.3464	0.01364
0.45	0.01772	0.2598	0.01023	0.3897	0.01534
0.50	0.01989	0.2887	0.01137	0.4330	0.01705
0.60	0.02362	0.3464	0.01364	0.5195	0.02046
0.70	0.02756	0.4041	0.01591	0.6062	0.02387
0.80	0.03150	0.4619	0.01818	0.6928	0.02728
1.00	0.03937	0.5774	0.02273	0.8660	0.03410
1.25	0.04921	0.7217	0.02841	1.0815	0.04262
1.50	0.05906	0.8660	0.03410	1.2990	0.05114
1.75	0.06890	1.0104	0.03978	1.5155	0.05967
2.00	0.07874	1.1547	0.04646	1.7321	0.06819
2.50	0.09843	1.4434	0.05688	2.1651	0.08524
3.00	0.11811	1.7321	0.06819	2.5981	0.10229
3.50	0.13780	2.0207	0.07956	3.0311	0.11933
4.00	0.15748	2.3094	0.09092	3.4641	0.13638
4.50	0.17717	2.5981	0.10229	3.8971	0.15343
5.00	0.19685	2.8868	0.11365	4.3301	0.17048
5.50	0.21654	3.1754	0.12602	4.7631	0.18753
6.00	0.23622	3.4641	0.13638	5.1962	0.20457

the micrometer should be if the thread is the correct finish size, use one of the following formulas.

For Unified threads $\quad M = (D + 3W) - (1.5155 \div P)$

where M = Measurement over the best wire size, in inches

D = Outside diameter of the thread, in inches

W = Diameter of the best wire size

P = Pitch (number of threads per inch).

For Metric threads $\quad M = P + C$

where M = Measurement over the best wire size, in millimeters

P = Pitch, in millimeters

C = Constant (from *Table 10.2*).

Note that *Table 10.2* has pitch, best wire size, and the constant stated in millimeters and converted to decimal inches, so M can be determined using the inch conversions in the table. When the conversions are used, the value of M will be in inches, rather than millimeters.

Forming threads with a die head

The use of a die head to form external threads is typically reserved for times when high production numbers are anticipated. In most applications, the die head is used on speed or turret lathes where the turret head can be indexed very quickly to position the tool. The unique feature of the die head is how the cutting dies must be *cocked* into a closed position so that when the threads are cut, and no more forward motion is applied to the unit, the dies spring open so that the cutting edges will clear the diameter of the workpiece for easy removal.

The actual setup of the die head is straightforward. Select the dies (typically four) by thread pitch and diameter and install them into the head. During the assembly, be sure to insert the dies in the proper sequence. Each die head will be numbered as to installation position, e.g., 1, 2, 3, and 4.

The operational procedure is not complicated, but it must be performed with some deliberation. You will have to develop a certain amount of feel to operate the unit successfully. This is especially true with the typical lathe because the tool will be fed into the workpiece via the tail stock feed wheel. On speed and turret lathes, the tool is also fed with a feed wheel, but the location and size of the wheel makes the operation of the die head much easier.

If you push the die head too fast, the threads will be over or undercut. In effect, the die head will act like a typical cutting tool and cut the workpiece to a certain diameter, leaving behind no threads, only a ragged surface finish where threads were supposed to be. When the die head is not fed fast enough, the automatic retraction action is engaged and the dies will spring back and out of the way of the workpiece.

Internal threads

Internal threads can be cut on a lathe with a tap installed in a tap handle, with an internal thread cutting tool, or with a tapping head.

Review Questions for Chapter 10

1. What two major lathe components are affected by backlash?
   A. Head and tail stocks
   B. Compound- and cross-slides
   C. Head stock and compound-slide
   D. Tail stock and cross-slide.

2. Facing operations are performed by pushing or pulling the cutting tool in or towards the workpiece _____.
   A. Centerline
   B. Length
   C. Diameter
   D. None of these.

3. Facing is done along the length of the workpiece.
   True or False?

4. Turning is done across the face of the workpiece.
   True or False?

5. The compound-slide feed dial will cut _____ or _____.

   A. 1:1 or 1:2

   B. 1:1 or 2:1

   C. 1:1 or 4:2:1

   D. None of these.

6. Offset turning is performed by _____.

   A. Moving the offset diameter to the head stock centerline

   B. Moving the offset diameter away from the head stock centerline

   C. Machining the part using the taper attachment

   D. None of these.

7. The compound will be used to cut a taper onto a workpiece, the slide positioned at 12°, what angle will the taper on the part be?

   A. 6°

   B. 12°

   C. 18°

   D. 24°.

8. For most efficient parting, the cutting tool should be placed at either ____ or ____ degrees.

   A. 89° or 91°

   B. 44° or 46°

   C. 1° or 2°

   D. None of these.

9. The typical center drill has a ____ degree taper.

   A. 45°

   B. 60°

   C. 82°

   D. 90°.

10. Most grooving operations performed are either _____ or _____ grooves.

    A. V-cut or square cut

    B. V-cut or round cut

    C. Square or round cut

    D. None of these.

11. When external grooving, the cutting tool should _____ _____.

   A. Be held perpendicular to the workpiece
   B. Not be overextended
   C. Both A and B
   D. Neither A nor B.

12. What is the most common thread pitch angle?

   A. 30°
   B. 50°
   C. 60°
   D. 10°.

13. What are the two most common ways to measure threads?

   A. Caliper and micrometer
   B. Caliper and setting fixture
   C. Go/No Go gage and with wires
   D. None of these.

CHAPTER 11
Project Planning

Process Sheets

Planning and preparing for a lathe project is very important—you cannot just step up to the lathe, turn on the switch, and start machining. So far, we have examined a great many procedures used to prepare the lathe for work: now it's time to learn how to coordinate a job.

When preparing for a trip, you will look at maps, make transportation arrangements, book hotel accommodations, plan to see the sights, etc. The same amount of work is needed to prepare for a machining project. In this case, however, you will need to look at the drawing or blueprint for specifications, develop a process sheet, acquire the material for the job, locate and purchase the tooling, and select workholding tooling. Finally, you will be ready to perform the work.

Your first task will be to get the part specifications. This is accomplished with a set of blueprints or field drawings. Blueprints are formal drawings

Figure 11.1 This machinist is laying out his tools in an organized and logical fashion to reduce wasted time and increase his production rate.

generated by hand or with a computer (sometimes called computer aided drafting or CAD) that contain specifications and information necessary for you to complete the project. Items include material type, number of pieces, surface finish, heat treatment, and tolerances to describe the part's size, hole locations, chamfers, etc.

The creation and use of a process sheet will speed up the completion time of any job. Learning how to think a job through, visualize machine moves, and how to select tooling will enhance your machining skills and proficiency. Process sheets allow you to make lists of machines, tools, and procedural steps for the project, and they can be saved for future reference should you have to perform the same job again. Often, projects are very similar, so that the process sheet from one job can be used or modified for another.

The process sheet will contain several pieces of information, including the name of the part, material type, size of the stock to be used, when the sheet was created, a list of the necessary machines to complete the project, a list of tooling, and a list of the procedural steps with tool identification, and speeds and feeds for each step.

It does take some time to create a process sheet; but you will save time if you don't have to search for tooling or try to decide which step should be next. In other words, these sheets save time, and saving time will earn you more money.

Machine Lists

Making a list of machines will help you organize workflow in the shop. In busy shops, certain machines may be allocated for particular jobs or operations. This makes scheduling of that machine a part of your machining sequence priority. The project may require several machines in order to complete the number of operations necessary for job completion.

Your machine list should be laid out using a logical progression of sequences. Don't list the cut-off saw last if it is the first machine used to cut the stock to size.

Learn to develop a code, or shorthand, when listing the machines, e.g., machine #1, the cut-off saw, is denoted M1; #2 the horizontal mill, is M2; #3 the surface grinder, is M3, etc. You may want to make a list of all of the machines used in the shop and assign a machine number to each of them. This will cut down on confusion and save time. When your shop has more than one machine, such as a lathe, list them with a subcharacter, e.g., M4a is the 9" × 24" lathe, M4b is the 14" × 36" lathe, etc. Tooling can be identified with a T code number, e.g., T1 is the #3 center drill, T2 is the #7 drill, T3 is the $1/4$-20 plug tap, etc.

Tooling Lists

This list should include all of the tools necessary for the job, including hand tools, holders, cutters, fixtures, and clamping devices. Many machinists also include the hand tools necessary for a particular job or operation. This is because the majority of machinists have their own tools to work with and take them from machine to machine as they move around the shop. In those shops where tooling is kept in a tool crib and must be checked out, a list is definitely in order.

Be specific about the tooling list. After studying the prints, make a list of those tools necessary to cut, drill, ream, and bore the part. Decide which type of holding or fixturing would be best suited for the machining process. If the part is one that is made frequently, there may be a specially designed holding fixture already made—check to find out. For those parts requiring standard tooling (chuck, quick-change toolholders, etc.), decide which ones will make the job easiest to perform. Remember, time is money!

Indicate the size and types of cutting tools necessary. Specify tool material type (HSS, carbide, coatings, etc.), shape, size, etc. Toolholders must be identified and sequenced. Differentiate which holders are used for quick-and-dirty jobs (those with loose specifications) and those that require a higher degree of precision. You wouldn't want to use a drill chuck to hold tools if a collet will offer more precise location of the tool bit.

Process/Step Numbers

The process or step should be numbered starting with the first thing you need to do. Typically, this would be to get the stock that will be machined. This may seem like a ridiculous place to start, but it isn't. More and more companies are using statistical process control (SPC) and ISO-9000 standards by which they are able to follow the progression of a workpiece from initial purchase of the stock, through machining, and finally to a completed product. As manufacturing methods become more standardized throughout the world, SPC and ISO will become the norm by which all products are produced.

SPC is a wonderful tool for the machinist if used properly. By tracking each part you produce, you will begin to see evidence of when your tooling starts to wear out. This will prevent overusing a tool, which results in a mistake and the scrapping of the part. Recording part dimensions may seem like a waste or misuse of time, but it's not.

The actual list of steps and the information you include in your process sheet may be driven by SPC or ISO standards, or you may be able to include a bare-bones list. Always make a list of the steps as company procedures dictate.

Many of the steps will describe a cutting routine, maybe with a drill bit, boring tool, and/or threading tool. Regardless, you will need to compute speeds and feeds for each one of them. Knowing how to compute the correct speeds and feeds will save money and aggravation.

Finally, you should include a final inspection sequence on your list. Production machine shops employ inspectors to perform routine and random inspections. Parts may be inspected in batches or as a complete lot. Typically, you will need to produce a first article for inspection before you are allowed to jump into the middle of a long production run. The first article would be the first part you produce once all of the tooling and machining operations are refined and set up.

Small shops will generally perform inspections in-house. One person may be designated as the inspector and it will be up to him or her to check parts for size, fit, and finish. Their decision will be final.

You will be the first person to check your own work. Because you are making the parts, initial checking of the parts will be your responsibility, so be sure that your measurement tools are in calibration by having them checked prior to starting the job. If your micrometer was .001" off, and the specifications gave a tolerance of ±.0003", you would have a lot of explaining to do to your boss when all or most of your parts were not produced to specification.

Learning how to create a process and setup sheet will also prepare you for advanced machining projects using computer numerical control (CNC) machines. These machines use a programming language unique to their application. If you understand the thought process required for the process sheets, it will be easier to understand the programming.

Process and Setup Sheet

Page ___ of ___

Part Name: _____ Date: ___/___/___

Material Type/Alloy: _____ Size: _____ Heat Treatment: _____

Surface Finish: _____ Ra

Machines:

M-1 _____ M-4 _____

M-2 _____ M-5 _____

M-3 _____ M-6 _____

Tooling:

T-1 _____ T-6 _____

T-2 _____ T-7 _____

T-3 _____ T-8 _____

T-4 _____ T-9 _____

T-5 _____ T-10 _____

Process/Step	Tool	Speed	Feed
1			
2			
3			
4			
5			
6			
7			
8			
9			
10			

Machinist: _____

Notes:

Review Questions Answer Key

Chapter 1

1. D
2. D
3. A
4. D. You are responsible!
5. B
6. D
7. A
8. C
9. A

Chapter 2

1. C
2. B
3. A
4. A
5. B
6. C
7. C
8. B
9. A

Chapter 3

1. C
2. B
3. A
4. B
5. A
6. B
7. D
8. A
9. D
10. B
11. C
12. D

Chapter 4

1. D
2. A
3. C
4. B
5. B
6. A

Chapter 5

1. D
2. D
3. D
4. A
5. B

Chapter 6

1. C
2. A
3. C
4. B
5. D
6. D
7. A
8. B
9. A
10. D
11. A
12. C
13. B
14. A

Chapter 7

1. C
2. C
3. A
4. B

Chapter 8

1. C

2. D

3. C

4. D

5. B

6.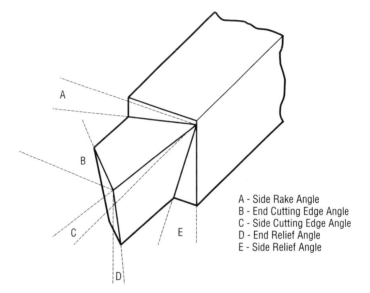

7. C

8.

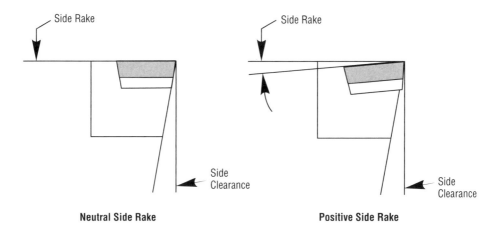

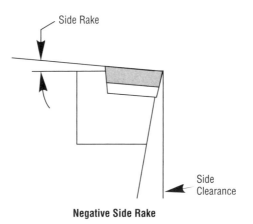

9. D

Chapter 9

1. B
2. B
3. D
4. A
5. B
6. A
7. C
8. B

Chapter 10

1. B
2. A
3. False
4. False
5. B
6. B
7. D
8. D
9. B
10. C
11. C
12. C
13. C

INDEX

alignment, 160
angle of attack, 60, 184
angle of inclination, 178
apron, 41
attack angle, 60, 184
axial cutting force, 183
axial runout, 159

back rake angle, 178
back-gear, 36
backlash, 211
ball-screws, 213
bed, 39
bell type center drill, 227
belt drive system, 35
boring tools, 101-107
brazed point cutting tools, 84-94
bull-head center, 131

calibrated scale, 47
carbide tools, 81, 84
carriage, 41
carriage movement indicators, 51
center drill driver, 77
center drilling, 227
center drills, 99, 227
centers, 127
    bull-head, 131
    dead, 127
    live, 127
chip thickness, 184-184
chuck, balancing limits, 136
chuck, gripping power, 135
chuck key, 74, 134, 142
chuck, maximum RPM, 136
chuck safety, 143
chucker lathe, 145
chucks, 133-147
    2-jaw, 127

chucks, *(continued)*
    3-jaw, 133, 137-139
    4-jaw, 140
    5-C collet, 144-147
    6-jaw, 141
    drill, 69
    keyless, 73
    rotating, 142
    self-centering, 139
    universal, 133, 137-139
clearance angle, 181
collet closers, 147
collets, 144-147
compound-rest, 44, 49, 149
compound-slide, 44
computer aided drafting (CAD), 246
computer numerical control (CNC), 30, 75
continuous cut, 171
coolant, 197, 202-204
coolant application, 18, 198
coolant disposal, 203, 204
cross-slide, 43, 48, 49, 148
cutting feed (ipm), 174-176
cutting fluid application, 198
cutting fluids, 17, 197
cutting forces, 183
cutting speed (sfm), 173, 174
cutting speeds, recommended, 177
cutting tools, boring tools, 101-107
    brazed point, 84-94
    carbide, 81, 84
    form tools, 108
    geometry, 178-186
    high-speed steel (HSS), 81, 83
    indexable insert, 94-96
    knurling tools, 109
    maintenance of, 111

cutting tools, boring tools, (continued)
  radius cutters, 108
  single-point, 83, 84-94
cutting-off, 226-227

dead center, 127
depth control attachment, 30
depth of cut (d.o.c.), 176
dial indicator, 161-165
die head, 238
digital readout (DRO), 51, 161
dog, drive, 129-131
dovetail way, 43
drill bits, 96
drill chuck, 69
drills, center, 99
  center, driver, 77
  jobber length, 96, 97
  Morse taper, 98
  screw machine length, 96, 97
  taper length, 96, 97
drive dog, 129-131

ear protection, 3
electrical requirements for lathe, 32
end cutting edge angle, 181
end relief angle, 181
external grooving, 228
external threading, 232
eye protection, 2

face plate, 126
facing, 213-218
feed ratio, 48
5-C collet chuck, 144-147
flood cooling, 17
follower rest, 151
foot stock, 44
form tools, 108
4-jaw chuck, 140

gap face plate, 126
gap-bed lathe, 29
gear drive system, 35
gib, 20, 43
Go/No-Go thread gage, 234
graduated scales, 48
grinder, tool post, 110
grooving, 228-230

half-moon rocker, 60, 61
hardness, Rockwell, 82
head stock, 32, 45
high-speed steel (HSS) tools, 81, 83

inches per minute (ipm), 174-176
indexable insert boring tools, 101-107
indexable insert toolholder, 74
indexable insert tooling, 94-96
indicator, dial, 161-165
indicator, test, 162
insert geometry, 179
internal grooving, 230
internal threading, 239
International Organization for Standardization (ISO), 249
interrupted cut, 171, 172

Jacob's taper, 74
jobber length drill, 96, 97

key, chuck, 74, 134, 142
keyless chuck, 73
knurling tool, 109

lathe components, 27
lathe construction, 31
lathe dimensions, 26, 29
lead angle, 181, 184, 186
left hand tooling, 76
level, spirit (bubble), 166

leveling the lathe, 19
live center, 127
lubrication of lathe, 13

machinability ratings, 186-193
machine lists, 247
magnetic swarf removal, 205
mandrels, 151
manual-speed change, 35
Material Safety Data Sheets (MSDS), 6, 203
micrometer stop, 51
Morse taper, 46, 69, 70, 127
Morse taper drill, 98

negative insert, 179

offset turning, 221-223
oil skimmer, 204, 205
oil, way, 14
On-Off switch, 33
over wires thread measurement, 234-238

parting, 226-227
personal safety, 6
positive insert, 179
process sheets, 245

quick-change gearbox, 39
quick-change toolholder, 67

radial cutting force, 183
radial runout, 159
radius cutters, 108
radius type center drill, 227
ram, 45
ratio (feed) 48
reamers, 100
rest, compound, 44, 49, 149
rest, follower, 151
rest, steady, 149

reversible jaws, 138
ride-along indicator, 53
right hand tooling, 76
Rockwell hardness, 82
rotating chuck, 142
round-cut groove, 228
runout, 159

saddle, 41
safety, personal, 6
screw machine length drill, 96, 97
self-centering chuck, 139
setup sheet, 251
side cutting edge angle, 181, 184
side rake angle, 180
side relief angle, 181
single-point boring tools, 101-107
single-point cutting tools, 83, 84-94
6-jaw chuck, 141
speed lathe, 30
spindle, 36
    bore diameter, 37
    controls, 34
spindle nose, cam lock, 36, 122-124
    long taper key, 36, 124-126
    short taper, 36, 119-122
    threaded, 36, 116-119
square-cut groove, 228
statistical process control (SPC), 248
steady rest, 149
straight turning, 218-221
surface feet per minute (sfm), 173, 174
swarf removal, 205
swing, 26, 29

tail stock, 44, 68
tail stock indexer, 69
tail stock offset, 46
tangential cutting force, 183

taper attachment, 152
taper length drill, 96, 97
taper turning, 223-225
tapers, 223
taps, 100
test indicator, 162
thread measurement, 234
thread nomenclature, 231
threading, 230-234, 238
threading dial, 232
3-jaw chuck, 133, 137-139
tool geometry, 178-186
tool post grinder, 110
tool posts, 60
    four-way, 62
    open-side, 61
    rocker base, 61
    strap-and-stud, 62
toolholder dimensions, 64-66
toolholder, indexable insert, 74
toolholder, quick-change, 67
toolholders, 59, 63, 64

tooling lists, 248
top jaws, 137
top rake angle, 178, 179, 181
tramp oil skimmer, 204, 205
troubleshooting, 20
turning, 218-225
    offset, 221-223
    straight, 218-221
    taper, 223-225
turret lathe, 29

universal chuck, 133, 137-139

variable-speed change, 35
V-cut groove, 228
vibration, 4

way oil, 14
ways, 39, 43
workpiece machinability ratings,
    186-193